A WINDOW IN THE SKY

A WINDOW IN THE SKY

Astronomy from Beyond the Earth's Atmosphere

A.T. Lawton

PERGAMON PRESS

New York   Oxford   Toronto   Sydney   Frankfurt   Paris

Library of Congress Catalog Card Number 79-65567

Photoset by Advertiser Printers (Newton Abbot) Ltd
and printed in Great Britain
by Redwood Burn Ltd, Trowbridge
for David & Charles (Publishers) Limited
Brunel House Newton Abbot Devon

ISBN 0 08 024663 X

Published in the United States of America
by Pergamon Press, Inc.
Elmsford, New York 10523

Contents

Acknowledgements

Many people have assisted me in the writing of this book. At the risk of causing offence through accidental omission:

I would like to thank Dr Robert Bradford, Dr John Best and Mr John Griffiths, all of EMI, who gave much by way of spare time, useful information and good counsel.

Also I would like to thank Professor Van de Kamp for discussion on the movements of Epsilon Eridani and Barnard's Star. Likewise, I must thank Dr John Billingham for discussion on the possibility of searching for extraterrestrial intelligence.

I acknowledge with gratitude the useful assistance given by Kenneth Gatland and Adrian Berry, in discussions, and by Penny Wright, in reading the manuscript and offering useful advice.

I am indebted to Paul Barnett of David & Charles for his unremitting patience in dealing with the manuscript.

Above all I sincerely thank my wife Āine for her continued encouragement and monumental effort in typing the manuscript.

A.T. Lawton
Shepperton
April 1979

A Note on Units

The units used in the book conform to a 'commonsense metric' scheme, rather than inflict the full rigours of the *Système Internationale* on the unsuspecting reader. In case some of the units are unfamiliar, they are as follows:

Distance. Millimetre, centimetre (1cm = 10mm = 0.3937in), metre (1m = 100cm), kilometre (1km = 1,000m), Astronomical Unit (1AU = 150,000,000km (approx.) = the distance of the Earth from the Sun (approx.)), parsec (1pc = 31,000,000,000,000km (approx.)). Where appropriate, rather than the parsec, the light year is used as a measure of distance (1pc = 3.26ly (approx.)); the light year is of course the distance covered by light (which travels at about 300,000km per second in free space) during the course of one year.

Temperature. Temperatures are given in Kelvins. Kelvins are of the same 'size' as Celsius (Centigrade) degrees, but their zero is not the melting point of ice but Absolute Zero, which corresponds on the Celsius scale to a temperature of approximately -273^0; i.e., $0K = -273^0C$.

Mass. Weights, masses and forces are given in grams, kilograms (1kg = 1,000g = 2.205lb) and tonnes (1 tonne = 1,000kg).

Angles. Angular measurements are given in degrees, minutes and seconds of arc ($1^0 = 60'$, $1' = 60''$); where appropriate radians are used (360^0 (a full circle) = 2π rad; 1rad = 57.3^0 (approx.)).

Large numbers are given as multiples of the powers of 10. 10^1 = 10, 10^2 = 100, 10^3 = 1,000, 10^4 = 10,000, etc.; this is because, say, 10^{29} is far less cumbersome than 100,000,000,000,000,000,000, 000,000,000. Thus 2×10^3 = 2,000, 2.36×10^4 = 23,600 and so forth. Very small numbers are given as the powers 10^{-1}, 10^{-2}, 10^{-3} ... where 10^{-1} = 0.1, 10^{-2} = 0.01, 10^{-3} = 0.01, and so forth.

Prologue

With satellites orbiting above the atmosphere and the prospects of men being in space for extended periods — years rather than weeks — our outlook is rapidly changing, as are our ideas as to how the Universe came into being and as to how it may die. We are in the position of someone who has lived in a cave all his life and who is now making a decision to come out and explore, to look at the big world outside, perhaps even to stay out there.

If we venture out, it will not be easy. We shall make new and remarkable discoveries — but many of us will die 'out of time and season' in strange, lonely and far-off places. For every new advance that is made, either in our understanding or in terms of our physical journeying, there will be a thousand failures.

The twenty-first and twenty-second centuries in space will probably be very similar to the sixteenth, seventeenth and eighteenth centuries on Earth: they will be periods of exploration and the gathering of knowledge. The succeeding centuries, up to, say, the twenty-fifth, *could* see the launching of ships to the stars, a follow-on from the unmanned interstellar flights made during or before the twenty-third century. When Man has spread among the nearer stars it is unlikely that he will ever face extinction, even when, in four billion years' time or so, the Earth is engulfed by the giant red Sun. Instead he could be in the system of, say, Tau Ceti watching the event *via* a satellite left behind in the Solar System, seeing the curtain rung down as the stage on which he has performed for millions of years is consumed by the flames. Now there would be no catastrophe large enough to threaten his existence until the Universe itself began to fade away into stillness . . .

But Man has to say 'yes' to the opportunity.

If he says 'no' then he can close the window in the sky, avert his eyes from what can be seen through it, and turn back to the warmth of his cave. The darkness, as Byron said, will then be his Universe forever.

1 The Invisible Barrier

All life on Earth ultimately owes its existence and well-being to a shell of cool moist air termed the 'atmosphere' (derived from the Greek *atmos*, vapour). Although an essential support to life, it is also a barrier. It protects the Earth and its inhabitants from lethal radiation emanating from the Sun and other sources elsewhere in the Universe, and it shields us from all but the largest of meteoritic bodies.

This protective shell is extremely thin, the main body being about 160km deep, but it becomes progressively less dense with increasing height; very faint traces have been discovered 8,000km above the Earth's surface. However, the 'invisible barrier' that concerns us is much lower. Over 99% of the Earth's population lives below a height of 3,000 metres. The diameter of the Earth is 12,750km and, therefore, our atmospheric blanket is a mere 1/4,000th of the Earth's diameter. To illustrate this with a small-scale practical example, it corresponds to the thickness of the wrapping tissue on a 15cm-diameter grapefruit.

We can take this small-scale analogy even further, for, if the tissue paper is of good quality, it will be semitransparent, and we should be able to see the colour of the grapefruit and any large marks or scars that there might be on the outer rind; however, we would not be able to see the tiny pores of the rind. As we shall see later, this is how the atmosphere behaves at certain wavelengths of the electromagnetic spectrum. At other wavelengths, however, the atmosphere behaves like aluminium foil on our grapefruit. We cannot see anything: the wrapping is completely opaque. And, finally, over a very select band of wavelengths, our atmosphere behaves like a film of transparent plastic — we could see almost every minute detail of the grapefruit: even the skin pores would be clearly visible.

But, transparent or opaque, our atmosphere in some respects is superior to the finest armour available. On 10 August 1972, a meteor estimated as being 4m in diameter and weighing 1,000 tonnes ploughed through the upper atmosphere over the western United States and Canada. At the lowest part of its trajectory it was 56km above Salt Lake City, and at this stage it literally bounced off the thicker layers and ricocheted back into space. What the atmosphere does for meteorites like this one it does also for gamma rays, X-rays and some sections of ultraviolet, infrared and radio wavelengths.

There may have been times in the past when the armour failed, due to outside events in the cosmos, and this may explain some twists to the key of evolution.

A Revealed Rainbow — The Electromagnetic Spectrum
The atmosphere is transparent to visible light and to radio waves. What do we mean by 'visible light'? What is the connection between such diverse phenomena as light and radio? We can see the effects of one with our eyes, but the other requires elaborate and complex receiving gear and even then we do not 'see' in the accepted manner; we have to display the information on a pen recorder chart or a television cathode-ray tube.

The answer to the first question was given by Isaac Newton, who in 1666 carried out a fundamental experiment, that of splitting white light into a bright band of colours: red, orange, yellow, green, blue, indigo, violet — the familiar colours of the rainbow. With typical thoroughness Newton then showed that these colours could not themselves be split by a glass prism; therefore he termed them 'primary' colours. The complete set of colours he called the 'Sun's spectrum', and in further experiments he proved that every member of the set had to be present to produce 'white' light.

Long after Newton's death, many suspected that other 'colours' lay beyond both the red and the violet ends of the spectrum, but proof of their existence required suitable detecting instruments. In 1800 William Herschel (discoverer of Uranus in 1781) proved that the most intense heating effects of the solar spectrum lay beyond the deepest red. His detector was a thermometer with a blackened bulb. He named the radiation 'infra red light', from the Latin *infra*, below.

10

This stimulated the drive towards investigating the other end of the spectrum, and in 1801 Johann Ritter established that by projecting the solar spectrum onto strips of white paper impregnated with silver chloride — which blackens on exposure to sunlight — the greatest effect was obtained beyond the violet end. We call this ultraviolet light (Latin *ultra,* beyond).

Clerk Maxwell's *Treatise on Electricity and Magnetism* (1873) clearly established a connection between radiant heat, visible light and ultraviolet light. Not only that, the links of the equation were forged into a longer chain which connected electricity and magnetism. He also predicted that, as long as a changing magnetic field was involved, then 'light' in some form was created! The eye could not always see this 'light', but Maxwell showed quite conclusively that the seemingly diverse properties previously observed were all directly proportional to the frequency (rate of oscillation) of the electromagnetic field. He predicted that other wavelengths would be found corresponding to lower and higher frequencies, and time proved him right.

In 1887 Heinrich Hertz performed a series of crucial experiments that conclusively demonstrated the transmission and reception of what we now call radio waves (Latin *radius,* a ray); originally they were called 'Hertzian' waves, later 'wireless' waves. They had all the properties of light in that they could be reflected, refracted, focussed into narrow beams, and absorbed. Furthermore, they travelled at the speed of light — 299,792.5km/sec; but the frequency of electromagnetic oscillation was much lower and the wavelength correspondingly longer, by a factor of 10^9 (10 followed by 8 zeroes).

At the end of 1895 Wilhelm Roentgen serendipitously discovered X-rays while experimenting with a Crookes tube (the earliest forerunner of the modern television cathode-ray tube). Roentgen had modified his tube by inserting a small metal window at the positive or anode end of the tube. To screen out unwanted light he surrounded the tube with a cardboard shield, and found to his surprise that certain objects (notably the laboratory wall paint) glowed when the tube was switched on. Roentgen was very quick to note the importance of his discovery.

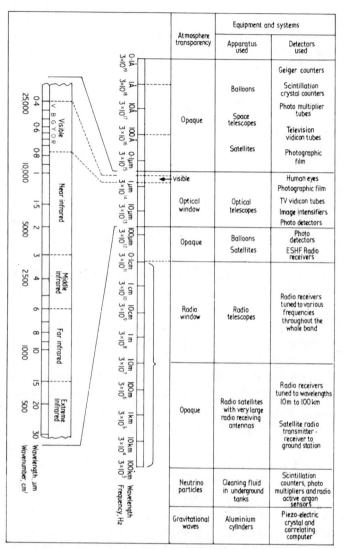

1 The electromagnetic spectrum, and the way in which it is explored in today's astronomy. The 'windows in the sky' show up in the optical and radio wavebands; otherwise the atmosphere is opaque, and we are forced to use high-altitude balloons and satellites. While not operating in the electromagnetic spectrum, neutrino and gravitational detectors have been included since they form a complementary part of modern astronomy.

The wall glowed green; the tube emitted a deep violet colour when unshielded. So radiation had somehow penetrated the cardboard, a fact quickly verified by placing such objects as 'thick books, stout deal boards, human flesh and other substances' in the expected path of the rays and at a distance of two metres. These too were penetrated and the rays shown to cast shadows on photographic plates 'precisely as they were by common or visible light'. (This use of the comparison 'light' in talking of what we now call 'X-rays' is taken from a contemporary account — October 1896 — and is a powerful indication of the impact of Maxwell's work.) When studied in detail X-rays were shown to be 1,000 times shorter in wavelength than violet light.

In 1896 Henri Becquerel discovered a further mysterious radiation with considerable penetrative power. We term this radiation gamma-rays, or γ-rays. Gamma-radiation has a very much shorter wavelength than X-rays, by a factor of about 10^9, and has greater penetrative power as a result.

Sadly, Maxwell himself did not live to see these far-reaching discoveries; he died in 1879 at the early age of 48. Had he lived, he would have seen the vindication of his Unified Electromagnetic Field Theories. (Mind you, had he lived another thirty years he would have seen this apparent smooth harmony almost utterly destroyed by the quantum physics of Planck, whose world was dominated by discrete quanta or packets of wavelength and energy and the maths of probability and uncertainty. But that, as Sheherazade said, is another story.)

When Maxwell came into this world the rainbow as understood by science was much as Noah knew it in *Genesis*: red, orange, yellow, green, blue, indigo, violet in scientific terms, a range of wavelengths extending from 7×10^{-7} metres down to 4×10^{-7} metres — not quite 2 to 1 in coverage. Ten years after Maxwell left this world, his revealed rainbow extended from long-wave radio at 10^4 metres down to γ-radiation at 10^{-16} metres. This is a range of 10^{20} — a truly astronomical ratio. If we represent the red and violet of Noah's rainbow by two points 3mm apart, then, on the same scale, Maxwell's rainbow would extend from Earth to a star nearly 10 parsecs distant! If that star were identical with the Sun we might have difficulty seeing it with the naked eye.

The Blanket has Holes

When the extent of this scale was realized, the true blanketing nature of the atmosphere became apparent. Marconi's practical use of radio seemed to confirm it, for the waves were actually bouncing back from reflecting layers (now collectively called the ionosphere) high above the Earth's surface. Not until 1931, when Karl Jansky first investigated the background noise of short-wave radio and proved that it came from the centre of the Galaxy, was it realized that a second window might exist.

In 1942 British radar sets were plagued by a sudden and alarming increase in background noise. This was traced to a large solar flare — and sowed the seeds of future radioastronomy investigation. The decade following the Second World War was largely a period of exploration, basically measuring the size and transparency of the radio window. This was combined with design and development engineering; the aim was to establish criteria for the building of equipments which were tools with known accuracy, performance and reliability; which were capable of carrying out a variety of tasks or being easily adapted to new work.

These years established the foundations of radioastronomy and showed that the radio window, in terms of wavelength, extended from about 20×10^{-3} metres (20mm) to about 20 metres. At the short-wave end it was limited by atmospheric absorption and at the other end by ionospheric reflection.

The radio window was thus about one hundred times as wide as the optical window. Moreover, it was found that, although the early designs did not have anything like the angular resolution of their optical counterparts, they were much more sensitive. Today (1978), the resolving power of a large single radiotelescope (typified by installations in the US and USSR) approaches that of an optical one, and, by using long baseline interferometry (a technique fully described on pages 109-13), a radiotelescope can resolve features presently inaccessible to the optical giants. A measure of the strides made in this field was the award of the Nobel prize to Sir Martin Ryle* in 1974 for 'advances in the techniques of accurate measurement in the field of radioastronomy'. The accuracy of angular resolution established by Ryle corresponds to the size of a postage stamp on the Moon as measured from the Earth's surface!

But still the dead hand of atmospheric absorption sits heavily on the band extending from 10^{-5} metres to 10^{-3} metres. Astronomers are now wishing access to this forbidden region, and lasers and masers are the tools to do the job. Lasers have now been built that operate at optical wavelengths from ultraviolet down to infrared, and masers from centimetric wavelengths up to fractions of a millimetre where they overlap lasers.

Millimetric radioastronomy should give the scientist details of bodies and regions of the sky that have temperatures around 100 to 200K. It is in these regions that 'warm' gas clouds may be condensing into stars and protoplanetary systems. Tantalizing glimpses of this phenomenon have been seen on photographic plates, but the major effects should be seen in the millimetric wavebands.

The best results will be obtained if apparatus working in this band is sited 'above the atmosphere'. The inverted commas are deliberate for, as we shall see, there is a choice in seeing without air. Each alternative has its advantages and snags, and all have to be weighed in the balance pan of the Exchequer, be it in dollars, pounds, marks, roubles or yen!

The 1960s and 1970s saw a real turning point, where further progress in astronomy demanded the placing of instrumented satellites in Earth orbit and beyond. They also saw developments in electronics, data processing and rocket propulsion that allowed such satellites to be placed in orbit at a cost which, although large, could be justified.

So, before we discuss instruments in space, perhaps we should look at what is presently our only known means of getting there.

*Now the Astronomer Royal.

The Rocket — Key to Pandora's Box

Briefly, a rocket motor develops thrust independent of its surroundings and is able to work underwater, in the air, and in the near vacuum of space. It does *not* work by 'pushing against the air of the atmosphere': despite several televized shots of the Apollo second stage firing (when it was about 50km up), and photographs of the Apollo Lunar Module taking off from the Moon in an almost perfect vacuum, this fallacy is still circulated.

A rocket works by thrusting against *itself*, and obeys Newton's Third Law of Motion: 'Action and Reaction are Equal and Opposite; for every action employed there is an opposing reaction.' The best everyday example of this is the movement of balls on a billiard table. The white ball strikes a coloured one and, depending on the bias given by the player, rebounds or stops. If the white ball rebounds smartly then the coloured ball glides slowly into the pocket. If the white ball almost stops then the coloured one slams into the pocket with almost all of the speed originally possessed by the white one.

Another example of action and reaction is the firing of a rifle. The bullet moves forward at a high velocity and the butt 'kicks' or jumps against the shoulder of the marksman. The size of this impact depends on three factors: the mass of the fired bullet, the velocity with which it leaves the rifle barrel, and the mass of the rifle. The product of mass and velocity is called 'momentum'. The firing of a rifle is a closed system; i.e., all the momentum available is shared between the speeding bullet and the recoil of the rifle. And it is shared 50-50; half goes to the bullet and half goes to the rifle. (The firer is not killed because the rifle is much more massive than the bullet.) No other forces are involved — hence the term 'closed system'.

The statement on the sharing out of momentum is not *quite* true. As the observant reader may note, I have forgotten the mass of the hot gas expelled with the bullet! Although gaseous, it still has mass and therefore possesses momentum. Let us imagine that we fire a 'bullet' of flour or face powder or similar material that will disintegrate as soon as it leaves the rifle. The kick from the butt is just the same, for the same mass is being expelled with the

16

same velocity. Now imagine a bullet that is all hot gas. Again the kick is unaltered. Finally, imagine that the rifle is a special — a hand-held multibarrelled affair, firing over 10,000 times a minute. If our indomitable marksman attempted to fire this device, he would not feel an intermittent kick but a straightforward continuous push which would probably thrust him off his feet. If our bruised but courageous marksman sat on a super-skateboard he would move off in the opposite direction to that of the rifle barrels. We have a crude rocket, the barrels generating thrust as they fire their charges of hot gas.

In reality what is termed a solid-fuel rocket motor is little more than this. Instead of a rifle barrel we have a tapered bell-mouthed nozzle which expands the hot burning gases in a controlled manner. As the exhaust gas expands, it cools down, and it increases its speed as it proceeds down the bell mouth. In engineering parlance it exchanges the potential energy of the relatively static but hot gas at the combustion chamber (the top of the bell) for the kinetic energy of the fast-rushing but cooler gases emerging from the nozzle of the bell mouth. The better we can make the expansion ratio and the hotter we can burn the gas, the higher will be the energy given to the gases emerging from the nozzle and the greater the thrust. This is analogous to firing our rifle bullets at a faster rate.

Early solid-fuel rocket motors used cordite as a propellant, but modern motors use synthetic-rubber-based materials as both fuel and binding material. The oxygen required for combustion of the fuel is contained in an oxygen-rich chemical such as ammonium perchlorate. The rubber-perchlorate is mixed to a smooth liquid and poured or cast into the motor body, and the nozzle assembly and ignitor system are attached after the mixture has set. Solid motors are noted for their extreme reliability and simplicity and are used extensively as boosters for satellite launch vehicles. Typical examples are as follows:

a) Aerobee — One of the first high-altitude research rockets.
b) Thor-Delta — A currently used satellite launching vehicle which may have up to nine solid-fuel strap-on boost motors

attached to the first stage, each with a thrust of 23,500kg.

c) Titan III M — A currently used satellite launch vehicle with two solid-fuel strap-on boost motors. Each motor develops 640,000kg of thrust.

d) The Space Shuttle — Currently the world's largest solid-fuel boost motors will be used on the Shuttle. Two such motors are ignited at take off. They are 45.4m long, 3.7m in diameter, and each develop a thrust of 1,202,000kg.

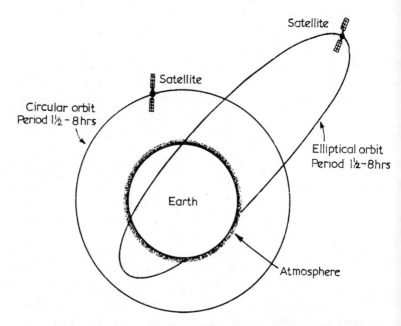

2 Simple orbits about Earth. If an object is propelled to a height of about 150km at a velocity of about 28kps it will describe an orbit in approximately one hour and twenty-five minutes. By increasing the velocity, the orbit is enlarged and, as one might expect, the period of the orbit increased. Corrective application of thrust can change the orbit into the pronounced ellipse shown here.

Note that we are now talking not of the sharp kick of a rifle, which is a force of a few kilograms, but of a long-duration thrust of approximately 1,200,000 tonnes generated for two minutes!

At the end of that time, the whole assembly will be at a height of 43km, whereupon the boosters will fall away and parachute back for recovery and re-use.

A second type of rocket motor which has a higher performance is fueled with liquid propellants. The price that is paid for better performance is increased complexity. In the case of the liquid-fuel motor, this complexity takes the form of tanks, pipes, control valves and turbine-driven pumps used to *force* fuel into the combustion chamber. The combustion chamber pressure of a modern liquid-fuel motor is usually very high — figures of 100-130 atmospheres are not uncommon — and these high pressures combined with high temperatures present complex cooling problems to the design engineer, for this high performance has to be contained in the lightest possible structure.

And, for good measure, the system has to have a proven reliability and life expectancy.

Earlier large liquid-fuel motors were usually fired only once in their working lifetime: other firings were test runs. The contemporary motor, as used in the Shuttle, is intended for reuse after recovery and check out, and may undergo up to fifty or more repeated firings (excluding tests).

The most powerful *molecular* chemical combinations that can be used in liquid fuel motors are:

a) *Liquid Hydrogen/ Liquid Oxygen.* This combination is the principal one used for high-performance motors and has been chosen for the Shuttle. Advantages are: known technology (proven in other projects), non-toxic exhaust (steam), and low-cost fuel commonly available and easily prepared.

b) *Liquid Hydrogen/Liquid Fluorine.* Although significantly more energetic than the oxygen combination, this system has not progressed beyond the experimental stage. The reasons are not hard to find. Liquid fluorine is not easily stored and is a toxic material to handle, as is the combustion product (hydrogen fluoride—HF). HF is also extremely corrosive, and when dissolved in water forms hydrofluoric acid — a well known glass-etching fluid.

Fluorine is also not as common as oxygen and therefore more costly to extract and liquefy.

Other combinations include liquid oxygen and hydrocarbons, notably kerosene.

The aim of the designer is to use the combination with the highest possible value of specific impulse (SI). This is a measure of the energy potential of the combination and is given in *units of thrust generated per unit of propellant consumed in unit time*. Thus one combination may produce 200kg of thrust per kg per second. This would be quoted as an SI of 200kg sec — or just '200'. Another combination *under identical conditions* may produce an SI of 300. Obviously the latter combination is almost twice as effective.

'Exotic fuels' have been theoretically discussed for many years but are not yet likely to be used in practice. They are based on radicals, sometimes termed monatomic fuels, and give *extra* energy when they assume a diatomic or molecular form. The most potent combination would be liquid triplet helium and monatomic hydrogen. However, there is no known practical means of preparing and storing these exotic materials in safety.

There is also, as something of a red herring, the nuclear-pulse rocket system as outlined in the 'Daedalus' study recently completed by a team of the British Interplanetary Society. The energy available is sufficient to place an astronomical observatory in the vicinity of another star system. Ironically, the ability to build and operate such a propulsion system may materialize quite rapidly in the near future, and possibly completely bypass the need for unstable monatomic chemical propellants.

Within the next 200 years we may be seeing close ups of 'sunspots' on Sirius or Barnard's Star, or examining other nearby objects which, although classed as stars, differ greatly from our Sun. We may even catch a glimpse of a planetary surface, such is the approaching sophistication of 'seeing above the atmosphere'.

The concept of landing a Viking spacecraft on Mars and

obtaining data from it might have seemed so utterly impossible to Newton or Maxwell as to be dismissed as absurd speculation, yet manmade astronomical stations orbiting in alien solar systems will follow the path navigated by Viking, Voyager and other probes that will extend our frontiers of knowledge.

2 Instruments in Orbit

In 1945 *Wireless World*, a UK technical radio magazine, published an article by Arthur C. Clarke which referred to the use of radio relay stations which could be placed in a 35,500km orbit. The significance of this distance was that the satellite rotated synchronously with the surface of the Earth at the Equator; indeed, it is now called a geosynchronous orbit, the satellite completing its circuit in 24 hours and thereby staying over the same spot on the Earth's surface.

Clarke has consistently performed better than most futurologists, and in this particular case he was right in his concept of using three such stations spaced at 120° intervals, but completely wrong as regards technical details. He visualized large structures housing maintenance crews working on a regular shift basis. The station therefore had to be equipped with a biological environment support system which included artificial gravity provided by rotating the whole structure. This auxiliary equipment also required a maintenance crew.

In a later discussion, Clarke pointed out how this situation arose. 'I based my ideas on the high-powered valves [vacuum tubes] then currently in use. They seldom lasted longer than 48 hours without burning out and I naturally assumed regular maintenance by engineering personnel. I did not dream of the electronics revolution that would permit the use of miniature reliable unmanned satellites.' This revolution in electronics started in 1948 — just three years after Clarke's article was published in *Wireless World*. The movement has recently gathered fresh impetus with the widespread introduction of silicon integrated circuits (SICs), which are described in detail in Chapter 6.

Another expert — this time Major General Walter Dornberger, who was primarily responsible for the development of the V2 — wrote in his diary for early 1944: 'I could imagine the eagerness with which meteorologists, physicists and *astronomers* would look forward to their first voyage into the stratosphere and ionosphere. . . We had often talked about the design and appearance of a space rocket: how windows giving an uninterrupted view from the spaceship could be made to keep off injurious ultraviolet rays from the Sun; how, in the chill of cosmic space, reasonable temperature might be maintained by means of black-and-white paint and the alternate turning towards the Sun of heat-absorbing and heat-reflecting surfaces.' This was written about six months before the first V2 crashed on London in September 1944.

Dornberger later went to the USA and, during 1947 – 1951, carried out a detailed study for Bell Aircraft on a project which can truthfully be called the ancestor of the Space Shuttle.

Both Clarke and Dornberger expected men to be placed in suborbital flight in the earliest phases of *any* projected space programme. Naturally, when these programmes came to be costed they were expensive — and governments were not prepared to fund money on a scale of typically 20 billion dollars (today's valuation).

The history of all present-day Western rocketry can be traced back directly to a paper produced by three engineers from the UK aircraft industry. It was read at the Second International Astronautics Federation Assembly (London, 1951). With the modest title of 'A Minimum Satellite Vehicle', by K. W. Gatland, A. W. Dixon and A. M. Kunesch, the paper showed quite conclusively that a useful payload of 10kg could be placed in low Earth orbit by a three-stage vehicle, the hardware of which was either in existence or required the minimum of modification. Definitely a 'beggar's satellite', MSV was evaluated as costing about 2% of the figures being quoted for more grandiose schemes.

MSV — or Mouse — was never built, but the USA fully acknowledges that this study laid the foundations for Vanguard and Explorer, the latter being the first US satellite in space (31 January 1958).

On 4 October 1957 the world awoke to hear beeping radio signals from a Russian satellite called Sputnik (Fellow

Traveller) which orbited the Earth every 96 minutes. It had an apogee (furthest point) of 947km above the planet's surface. The figure that shook the Western World was its weight: 83.6kg, over fifty times the projected weight of the Vanguard satellite's 1.6kg. For a while it was suspect.

These doubts about the launching capabilities of Soviet vehicles were quickly dispelled when Sputnik 2 was launched. It weighed 508kg and its apogee altitude was 1,670km; moreover, it carried a passenger, Laika the dog, the first creature in orbit. Further evidence of what unmanned vehicles could do was shown in the Luna (originally Lunik) series, especially when Luna 3, launched on 4 October 1959, passed round the far side of the Moon and successfully returned the first photographs of this hitherto inaccessible area.

Although crude by contemporary standards, they proved that unmanned vehicles could perform precision tasks of great complexity under the arduous conditions of free space, and that such tasks were well within the capabilities of then existing rocket boosters: they did not require the development of the colossal machines visualized by the manned vehicle protagonists. In the two decades following the launch of Sputnik, the Soviets have built up an impressive series of unmanned launches. There have been twenty-six Luna spacecraft — two of which carried Lunokhod roving vehicles and three of which returned Moon soil samples. In addition, ten Venus probes have been launched, two of which returned photographs of that planet's surface. Eight Zond spacecraft and seven Mars probes have been launched but can be regarded as only qualified successes. To date (May 1978) 1,004 Kosmos satellites have been launched and many are still in orbit — although one made headlines when in January 1978 it crashed on Canadian soil with sections of its nuclear reactor power generator still intact. Additionally, there have been several thousand suborbital flights, both military and meteorological, from the USSR.

The USA has been equally active, and has performed feats of remote computer controlled engineering which would have been regarded as near miraculous twenty years ago. Now, spacecraft are built to a closely controlled and costed programme and their reliability is highly predictable. It is no longer 'just good luck'.

Thus the Mouse triumphed, in the end, and development of high-reliability lightweight remote-controlled electronic equipment has exploded into the present-day multi-billion-dollar aerospace industry.

The 'Bullets' of Van Allen

The Russian success was made to appear as a political as well as a technical triumph, and the resultant potential damage to the Americans' prestige stung them into rushing the US Navy backed Project Vanguard onto the launch pad on 6 December 1957 — with disastrous results: the first stage exploded on the launch stand. A second launch, on 5 February 1958, also failed.

Ironically, the US could easily have beaten the Russians, and possibly launched a successful satellite in 1956! Backed by the army, Wernher von Braun had been busy further developing the V2 into a long-range ballistic missile, and, following the guidelines laid down in the paper by Gatland, Dixon and Kunesch, he had developed a satellite-launching system called 'Juno 1', which was a four-stage version of the existing Jupiter-C (a performance-improved V2). The US Government was not impressed by a scheme which, although American, had been designed by a German team based on work that originated in the UK. Funding was therefore cancelled, and the Vanguard, which then existed only on paper, was allowed to proceed, with the resultant humiliation of a widely publicized launch failure.

Finally, commonsense prevailed, and on 31 January 1958 the first Juno-1 launch successfully placed Explorer 1 into orbit. Almost immediately the modest satellite made an unexpected discovery. Part of the payload included a Geiger radiation counter used for experimental work by Dr James Van Allen of Iowa State University.* He was interested in cosmic particle

*Now with Ann Arbor University, Michigan.

intensities in the upper atmosphere. The counter functioned correctly during parts of the orbit, but was silent at other times. Originally Van Allen thought that this was failure due to an intermittent connection — the major headache of any electronic system. Then he realized that the 'failure' was cyclic, and he correctly deduced that the counter was silent because it was saturated (i.e., overloaded). As a result of further experiments with later Explorer *and* Vanguard satellites, he concluded that the Earth was surrounded by two toroidal (doughnut-ring-shaped) belts of charged particles.

The inner belt is at an average height of 3,000km and consists largely of protons, while the outer belt is at an average height of 22,000km and is composed mostly of electrons. These particles, which are trapped by the Earth's magnetic field, reach us as part of the Solar Wind, a high-speed stream of protons and electrons which is ejected radially in all directions from the outer corona of the Sun. It is propelled at about 800 to 1,000km per second by the impetus of the radiation pressure of sunlight. The Van Allen belts are the direct connection between solar flares (which increase the intensity of the solar wind) and such earthly phenomena as radio blackouts and aurorae.

Such a connection had been theorized by Carl Størmer, Sir Edward Appleton and others during the late 1920s and 1930s, but Explorer 1 provided experimental proof in a way which could not possibly be done by Earth-based equipment. We had to reach out above 'the blanket'.

Now that theories concerning the Van Allen belts have been refined, and accurate measurements of the magnetic fields of other planets obtained (see Chapter 10), it has become possible to predict — with varying success! — whether or not a planet possesses radiation belts. Theory indicated that Mercury, Venus and Mars would not, but that Jupiter would be surrounded by belts whose radiation intensity exceeded that of Earth. Measurements made by Pioneer 10 in December 1973 and Pioneer 11 a year later confirmed that the radiation around Jupiter was 10,000 times more powerful than terrestrial values. During part of the Pioneer 11 encounter the spacecraft flew ·

through regions of extreme intensity and some of the onboard electronics suffered temporary radiation damage even although the circuitry had been 'hardened'. (Radiation damage and hardening of electronics is covered in Chapter 5.)

I have described the Van Allen findings in some detail because this incident highlights how the unexpected is discovered. It also emphasizes the protective rôle played by the atmosphere, for in this case the effect is beneficial to us. If by some magic function we could survive the removal of the atmosphere, then life in the polar regions of the Earth would have to cope with a severe radiation hazard.

The Van Allen radiation forms an additional source of background 'noise'. If a satellite is to be placed in an orbit which entails it passing regularly through one or both belts, then any experiment which involves radiation counting has to allow for this effect when evaluating the results. Orbits involving manned flight are planned to minimize encounters with these areas in order to reduce danger to the astronauts.

The Magnetosphere

The Van Allen belts are part of a complex layered structure known as the magnetosphere; it is not truly spherical, but is forced into a tear-drop shape by the Solar Wind. The section facing the Sun carries a bow shockwave similar to that of a fast-moving ship; the section pointing away from the Sun streams out as a long tail to the tear drop. Since the magnetosphere of the Earth is composed of ionized particles similar to those of the ionosphere, it is not surprising that it is capable of bouncing or reflecting radio waves. However, the particle density is not high enough to allow current commercial exploitation, although it holds promise for the future. Most experimental work to date is associated with *radio* frequencies of 5 to 10kHz (1 Hertz = 1 cycle per second) — i.e., normally in the audio range — whereas commercial shortwave radio uses 5 to 30MHz — i.e., about 1,000 times higher in frequency.

Magnetospheric propagation does have one very important difference. A signal — given the right conditions — can be naturally amplified. The magnetosphere can be made to act as a gigantic vacuum tube amplifier of a type known as a 'travelling wave tube' (TWT — the derived nickname is obvious). To operate this system we need a source of energetic electrons (supplied by a hot cathode and accelerated by a high-voltage anode) and also a magnetic field which is of just the right intensity to cause the electrons to travel in a spiral path of correct radius on their journey to the anode. The time taken to complete *one turn* of the spiral corresponds to one cycle of the frequency we wish to amplify. If we feed in a signal of this frequency at what I will call the 'low-energy' end (at the cathode), the signal will gain energy from the main electron stream as it travels to the high-energy end (anode) where it can be removed. The dimensions of commercial TWTs allow them to amplify centimetric wavelengths — and they are extensively used in commercial communications satellites.

Translating this to the magnetosphere, there is a magnet — the Earth's magnetic field. We have a source of energetic electrons spiralling backward and forward between the north and south poles of the magnet at a certain frequency. Theoretically, then, if we insert a signal into one end of the spiralling particle zone, an amplified version should emerge at the other. Since the particles are going in both directions, amplification should occur both ways. All that remains is to choose the right frequencies and devise a suitable method of injecting the signal into this gigantic amplifier. Because the amplifier has such large dimensions, the optimum radio frequencies are very low and are typically 5 to 10kHz (they are called VLF — very low frequency — wavebands); the wavelengths are correspondingly very long — 30,000 to 60,000m. Comparing this with medium-wave radio wavelengths (typically 300m) and long-wave (typically 1,500m) gives an idea of the scaling involved.

However, if the Earth's scaling were different in that its magnetic field were (say) fifty times more intense, or the physical path length of the Van Allen belts and magnetosphere shorter, then the frequencies at which magnetosphere amplification could take place would fall within the normal broadcast radio bands.

The effect is being thoroughly investigated and in recent experiments, conducted over a 12,000km great-circle path between Alaska and New Zealand, appreciable amplification was obtained. The frequency used was 6.8kHz (44,000m wavelength) and, in radio language, 'a gain of 30dB was experienced'. This corresponds to an amplification of 1,000 times; thus a transmitter of 1kW using the effect has the same effect as a transmitter of 1MW operating over a normal radio path.

These experiments are being followed with interest, for VLF radio waves can penetrate both land and sea, and in the latter case can possibly be used for underwater communication with submarines.

Orbits to Order

Provided that the rocket described in Chapter 1 has sufficient fuel and energy, it will lift our satellite to whatever orbital height and speed is required, but there are some fundamental limits to the freedom of choice.

Our satellite payload must be lifted high enough to clear the atmosphere with a sufficient margin to allow time aloft to accomplish the task; if the satellite is below this critical height it will reenter the atmosphere and burn up in the denser layers lower down. The height for such a marginal orbit is about 150-200km and a satellite at this level may last for a few days to a few weeks before plunging back to destruction. The orbital period for such marginal cases is about 85 to 92 minutes, and measurement of this period and the rate at which it reduces gives an accurate indication of reentry date, time and position.

Imparting a greater speed to the satellite allows it to take up a higher orbit and, since the satellite has a greater distance to travel in circling the Earth, the orbital period is greater. Thus a satellite orbiting at a perigee (nearest point to Earth) of 1,150km will orbit in a time of approximately 108 minutes. Some specialized spacecraft — e.g., the USSR Molniya satellites —

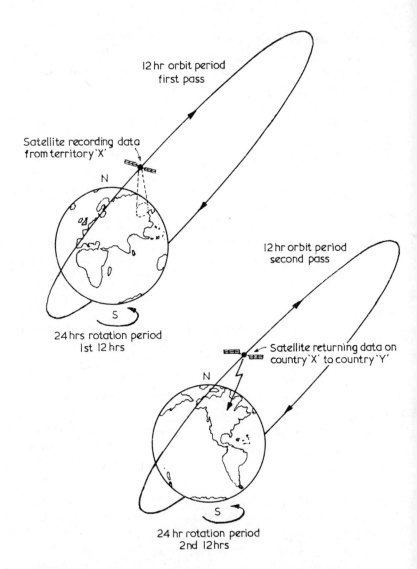

12 hr orbit period
first pass

Satellite recording data
from territory 'X'

N

24 hrs rotation period
1st 12 hrs

12 hr orbit period
second pass

Satellite returning data on
country 'X' to country 'Y'

N

24 hr rotation period
2nd 12hrs

3 Molniya orbits, so named for the Soviet system which originally
used them. The period of the orbit, which is highly elliptic, is 12
hours. During the first 12hr orbit the territory marked X is under
surveillance and may be suitably scanned by television or in the
infrared; during the second 12h orbit the territory beneath the
satellite is, owing to the Earth's rotation, 'friendly', and thus the
information gained during the first orbit can be beamed down for
analysis.

have highly elliptical orbits with perigee at about 430km, apogee (furthest point) at 40,750km, and an orbital period of 735 minutes (12.25 hours). This is a special polar orbit chosen to circle the Earth twice in one day. The high apogee allows the satellite to remain almost stationary above the surface and on the first pass it may be over (say) US territory. The low speed relative to ground allows photographic and television study in several wavebands. On the second high pass the satellite will be over the USSR and can transmit the acquired data back to the Earth-sited receiving station. It is fairly obvious that these items are generally used for military purposes.

Another interesting orbit is used by the US satellites ISEE-1 and ISEE-2. With a perigee of 340km and an apogee of 137,900km they have an orbital period of 3,440 minutes, which is 2.39 days. This high apogee reaches roughly one third of the distance from Earth to the Moon and allows the ISEE satellites to take regular samples of the conditions prevailing in the magnetosphere.

Escaping from Earth
If we impart still greater velocities to the satellites, they will eventually reach a point where they will not return to near Earth on a closed orbit. Instead they will 'fall past' the Earth and fly out into the Solar System. They will become satellites of the Sun, artificial minor planets in independent orbits. This critical speed is called escape velocity, and for the Earth it is 11.2km/sec. No matter what the size or weight of the satellite may be, it must reach this velocity if it is to break free of Earth's gravitational restraint.

The escape velocity of a body is dependent on its mass and radius. For example, the Moon has an escape velocity of only 2.38km/sec (the usefulness of this low value is mentioned in Chapter 4). On the other hand, the Sun has an escape velocity of 618km/sec — yet it loses mass at an average rate of approximately 635,000 tonnes per second due to the Solar Wind, a continuous stream of protons and electrons propelled to escape velocity by the pressure of the radiation emitted by the Sun.

31

Solar Wind effects can be measured by ISEE-3, the third member of the group, which was launched in 1978 and placed at what is called the Earth-Sun libration point. This is also termed a Trojan or Lagrange point — after Joseph Lagrange, an eighteenth-century mathematician-astronomer, who showed that a third body could form a semi-stable orbit with two existing bodies provided certain conditions were fulfilled. One of the conditions was that the third body must have a negligible mass compared with either of the other two. A further condition was that the third body must be in an orbit which has a similar distance from the primary as does the second, but it must occupy a position which is 60^0 ahead or astern of the second body. The whole grouping forms a giant equilateral triangle in space: in this instance the Sun, Earth and ISEE 3 occupy the three corners. Each, therefore, will be 150 million km from the other.

Several spacecraft are now orbiting the Sun as independent satellites. All of them move close to the solar equatorial plane in a similar manner to planetary orbits. The plane of the Earth's orbit is called the ecliptic and is inclined to the equatorial plane of the Sun by $7°$. The ecliptic is used as an orbital reference point for all the orbits of the planets and other bodies which make up the Solar System. The angular difference between the ecliptic and the plane of another body's orbit is called 'the angle of inclination', usually shortened to 'inclination'. Most of these angles are small; e.g., Mars $1.9°$, Jupiter $1.3°$, Uranus $0.8°$. Two planets, Mercury and Pluto, have high inclinations — $7°$ and $17.2°$ respectively — and the high figure for Pluto is one of the reasons for suspecting that it and its moon are not true planets but possibly escaped satellites of Neptune.

All of the planetary orbits, therefore, are confined to a very narrow slice about the equatorial plane of the Sun, rather like a thin layer of jam through the middle of a large sponge cake.

This is fortunate for existing spacecraft, for the energy content of present-day fuels is not sufficient to do much more than travel through the confines of this thin layer. The maximum achievable without further assistance is an orbital

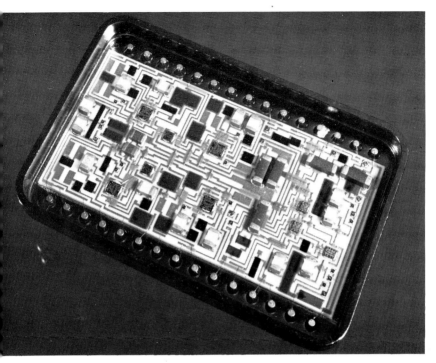

1 A multistage amplifier with nine silicon-chip assemblies, together with resistors, capacitors and diodes, shown immediately prior to capping (actual size). Note the thin wire bonds between circuit and connection pins. (*Courtesy EMI Electronics.*) See page 89.

2 A computer on a chip. A Ferranti F100L sixteen-bit fast microprocessor shown immediately before capping and sealing. Note the hairlike gold-bond wires (there are in fact forty) connecting the perimeter of the silicon chip to the surrounding strips. There are approximately one hundred thousand transistors and diodes in the chip assembly. The scale is twice full size. (*Photograph by Rex Londen; courtesy Ferranti Electronics.*) See page 84.

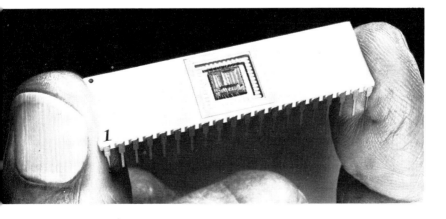

3 & 4 *Above*, a photograph of an EMI image intensifier with, on the right, a view of an encapsulated tube. The tube is 30cm long and provides an input photocathode and output phosphor diameter of 50mm. Its weight is approximately 1kg, but encapsulation raises this to 5–6kg. A diagram of the tube is shown *below*. The envelope is made from four similar stage sections, these being an assembly of two metal rings (3) with two metal flanges (5) at each end of the assembly, spaced and sealed together 25mm apart by three 75mm-diameter glass cylinders. To allow for the distribution of accelerating potentials within each stage two focussing rings (4) are mounted in each body. Targets (6) and (7), which comprise the phosphor photocathode sandwich on thin mica disks, are mounted between each stage section in the positions shown. Assembly of the tube is achieved by argon-arc welding around the periphery of the end flanges. The two end windows (1 & 2), of zinc crown glass sealed into metal disks, are likewise sealed to the tube by peripheral argon-arc welding. (*Courtesy EMI Electronics.*) See page 104.

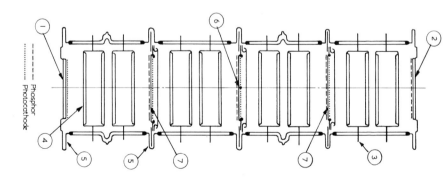

Phosphor
Photocathode

inclination to the ecliptic of about 20^0 at a distance from the Sun of 1AU.

But in 1983 a mission called OOE (Out of Ecliptic) or Solar/Polar will involve the launch of two 300kg spacecraft on a trajectory that will eventually take them 'over the top of the Sun'. To do this they will call on the massive gravitational slingshot effect of Jupiter. One of the craft will be flung over the north pole of the Sun while the other will travel in the opposite direction over the south pole. Slingshot orbits have already been perfected in the Pioneer 10 and 11 missions, which used Jupiter's gravitational field as a catapult to boost the velocity of the spacecraft such as to throw Pioneer 10 out of the Solar System altogether, while Pioneer 11 will pass close to Saturn in September 1979, approximately 6½ years after launch. Without Jupiter's assistance the mission to Saturn would have taken over twelve years and it is unlikely that the spacecraft would be working reliably after so long.

Fresh knowledge is expected from the OOE mission, which is targeted to last five years from the proposed launch date of 3 February 1983. The thin disk in which the planets reside is not typical of the environment surrounding the Sun. For example, our present view of sunspots at high solar latitude is grossly distorted. A polar view may help solve how sunspots are formed, for here is their birthplace.

Geostationary Astronomical Platform Systems

We have mentioned Clarke's proposal for communications satellites in geosynchronous or — to use the preferred term — geostationary orbits. His proposed three manned stations have been replaced by more than sixty unmanned items. Of these, forty-three are currently operational, the majority of them being run on a highly commercial basis. Clarke's 'pipedream' of 1945 earned a total of $1.8 *billion* direct revenue in 1975 for the US communications industry! The projected growth rate is estimated at not less than 11.8% over the next decade and *that* during a period of one of the worst recessions in US economic history. By 1985 it is estimated that satellites in the US private sector *alone* will earn an annual revenue of $5.5 billion.

There is even a situation of 'channel crowding' developing — a circumstance undreamed of in 1965, let alone 1945. To cope with this threatened communications explosion NASA are planning the development and use of a Geostationary Communications Platform (GCP). This is literally a giant automatic 'exchange in the sky'. Weighing 8.2 tonnes and requiring 20kW of power to operate it, the GCP represents a second-generation space satellite concept. Revolving once in 24 hours and keeping pace with Earth, one such satellite will handle the same traffic load as one hundred of our present generation of communications satellites.

A recently published paper by Fordyce, Jaffe and Hamilton of NASA details the economics of such structures and how they will be built. Three journeys by Space Shuttle are required to place the major subassemblies in low orbit. These are joined up to form the complete GCP, which is then carried up to the 35,200km geostationary orbit. After further testing, which includes setting the system to face accurately in the required direction, the GCP is left to look after itself, requiring servicing only once every five years.

I have described the GCP in some detail because in its construction, operation and maintenance lies the basis for a completely new generation of data-gathering astronomical instruments. I would call them Geostationary Astronomical Platform Systems (GAPS) although in certain cases it may be decided to place such platforms in orbits above or below the geostationary level.

One major advantage of such a platform would be its ability to transmit to Earth in real time the data acquired during observation. This stems from it revolving synchronously with Earth and therefore permanently hovering over the site of the receiving station. This cannot be done by ordinary satellites in low orbit which, because they revolve faster than Earth, usually rise in the west, set in the east and stay above the horizon only for short periods (15-30 minutes). This inability to transmit data directly in real time means that the low-orbit satellite has to store data on board and then transmit when it is in line of sight of a suitable receiving station. If there is only one station then the satellite has to give up the stored data at a proportionately faster rate than it may have been received.

For example, as discussed earlier, a low-orbit satellite may take 100 minutes to complete one revolution of Earth. If it is over the receiving station for 20 minutes then it must play back 100/20 — i.e., five — times as fast just to keep pace with the recording rate. Furthermore, it must store at least 80 minutes' and preferably the full 100 minutes' worth of orbital data. This represents an investment in reliable data storage — usually a high-precision tape recorder.

A GAPS, on the other hand, has no need to store data: it can play back as fast as it observes and processes. This means that it can use the full bandwidth available from the GAPS-to-Earth link on a complete 24-hour basis. In practice this will probably mean that several observational experiments can be carried out on the platform. An example would be simultaneous gamma-ray, X-ray, uv, optical, ir, and radio observation all from appropriate sensors trained on one sector of space. All of the signals would be multiplexed (i.e., time shared) together and transmitted down to Earth *via* a single link. Suitable back-up measures (e.g., a spare link) would be provided to guard against failures.

Simultaneous observation would be invaluable in survey work. If, for example, we could carry out a multiple-wavelength survey and discovered an area which produced radio and X-ray but not optical pulses we would be fairly sure that we had found a type of pulsar. Since the information content of several wavelengths is directly available we could probably immediately classify it.

Let us look at an example of the benefits of multiple-wavelength observations. Sir Bernard Lovell — Director of the Jodrell Bank Radio Observatory, and Professor of Radio Astronomy, Manchester University — has for many years studied the problem of receiving radio signals from flare stars. These are stars which suddenly increase in brightness by about

250 times (6 magnitudes) and then fade to their former level. Typically these increases in brightness last for a few minutes, but the actual flare-up itself takes only a few seconds. The true cause of the instability is not fully understood, but the stars themselves are believed to be young, rapidly rotating protostars which have not settled down to the Main Sequence. Flare stars are intrinsically very faint. They are sometimes known as 'UV Ceti' stars, after the prototype, and about thirty have so far been identified.

Lovell was interested in receiving the radio noise from true stars other than the Sun. The so-called radio stars are in reality objects such as galaxies, supernova remnants, quasars, pulsars, gas clouds and black holes, although the last have not yet been identified beyond all doubt.

Lovell reasoned that the radio energy created by a normal star was below the threshold of a typical large radiotelescope — with the exception of flare stars. Here the increase in brightness, if matched by a similar increase in radio brightness, would render it detectable by the Jodrell Bank installation. Observation of UV Ceti (a binary system made up of possibly the two smallest known stars), Proxima Centauri and several other flare stars proved that Lovell was correct: short-duration radio noise bursts *were* received. But the evidence had to be conclusive. This required the cooperation of optical observatories to monitor the candidate stars for visual signs of flares and to record the time, duration and magnitude of the activity. Correlation of the two sets of data then eliminated false radio signals due to interference generated locally. After years of work, Lovell was successful in his search, and he made the interesting discovery that radio noise was usually detected *before* the flares were seen.

The use of a GAPS would considerably simplify such an experiment. The monitoring of visual and radio channels would be easy, for they would automatically share a common timing basis. Visual observation would be improved because the GAPS was above the obscuring atmosphere, and the radio observations would benefit from the lower background noise of outer space, 35,200km away from manmade radio interference.

The system would give the bonus of ir observations as well, and it would probably be worth monitoring a flare star at uv wavelengths. Theoretically, one would expect the flare pulse to be first observed at radio wavelengths, then at ir as the temperature rose inside the flare disturbance. Eventually this would emit light, and therefore be 'seen', and if the temperature rose high enough it would be recorded in uv. The rate at which the flare increased and peaked through the various channels would provide a wealth of information which might provide convincing evidence for a correct explanation of 'flares'.

This highlights another advantage of the GAPS system: it can operate a 24-hour day of visual monitoring — an impossible task with an Earth-bound observatory. A space telescope in geostationary orbit would be prevented from working only if it were looking directly at the Sun (or Moon) or if it were obscured from its target by the bulk of the Earth; in geostationary orbit our planet would obscure about 20⁰ of the heavens. The use of two such stations would eliminate these problems and the use of three would allow several interesting experiments in long base-line interferometry (LBLI) to be carried out. LBLI (for a fuller explanation see pages 109-13) could be done at not only radio wavelengths but also ir, optical and possibly uv as well. This might yield direct observation of the diameters, 'sunspots' and other surface features of stars other than the Sun.

An idea of what to expect in this field of work was recently shown by computer-processed photographs of the red giant star Betelgeuse in Orion; this is the bright orange star that forms the hunter's left shoulder. When processing was finished the photograph showed huge 'sunspots' on the star's surface. The diameter of Betelgeuse varies between 200 and 400 times that of the Sun — placed in our Solar System this giant star would engulf Mercury, Venus, Earth and Mars, leaving just the other planets. The 'spots' are probably convection currents in the star's outer atmosphere, but they are of enormous size. The smallest detected is about 0.06 of the overall diameter of the star or roughly eighteen times larger than the Sun! It is believed that some stars may possess even larger surface blemishes, and GAPS sited in geostationary orbit should be uniquely placed to obtain high-resolution photographs of such features.

Clarke's dream is already a commercial success. Properly applied in astronomical fields it could reap an equally rich harvest of knowledge from the heavens.

3 Man in Space

Early proponents of spaceflight were convinced that Man would be an active participant in space journeys at the very start of such work. This was a natural extension of then current thinking. But, as we have seen, it was unmanned vehicles which made the first flights and Man did not venture above the atmospheric blanket until three and a half years later, when on 12 April 1961 the USSR announced the launching of Vostok (East). This made one orbit of Earth and carried Yuri Gagarin, the first man in space.

The majority of the manned flights subsequently made by both the USA and the USSR have been performed largely for political purposes, although much useful knowledge has been gained with regard to the overall response of the human system under space conditions.

The problem of manned vehicles is mainly one of size. Structures designed for long-term human life support *are* large. The Lunar Landing Modules and the Apollo Command Modules were comparatively small structures designed to accommodate two or three men for only a few days. The sheer size and scale of the Saturn 5 rocket system (weight 2,900 metric tonnes, first stage thrust 3,450 tonnes) reflected the energy and fuel requirements to place 51 tonnes of payload on the surface of the Moon. Had the vehicle been confined to Earth-orbit tasks, the Saturn 5 could have placed 152 tonnes in a 'low orbit' configuration.

There are only two types of manned vehicle currently or recently in use which are specifically intended for scientific research and which represent a true blend of men and instruments to form an integrated data-gathering system. They are:

The U.S. Skylab: Skylab, the third-stage cylinder of a Saturn 5, was modified and instrumented to accommodate three men for periods of up to 84 days. Weighing 76 tonnes, it was launched into a 435km near-circular orbit on 14 May 1973. It was abandoned on 8 February 1974, and fell back to Earth, scattering pieces over Western Australia, on 12 July 1979.

The USSR Salyut (Salute): This is the generic term given to a family of Soviet space stations, six of which have so far (1979) been launched. Salyuts 1-5 are believed to be mainly scientific in purpose; Salyuts 2 and 3 appeared to be semi-automatic military surveillance stations. Salyut 6 is the first of a new series, and its crews have carried out several unique experiments. In addition, they have several times broken the world endurance record: it is now over 140 days. Salyut 6 also performed the first three-vehicle and four-man link-up in space *via* a new double port docking system, and received food, instruments and fuel supplies from an unmanned automatic ferry vehicle — Progress 1. To round off this series of 'firsts', the Salyut 6 crew included the first astronaut not of US or USSR origin, the Czech Vladimir Remek. This is part of a multinational programme, for between now and 1983 there will be cosmonauts from Poland, East Germany, Bulgaria, Hungary, Cuba, Mongolia and Romania. Salyut 6 is so far the heaviest of the series, weighing in at 19 tonnes, the previous units being just over 18.5 tonnes.

The Soviet space stations are therefore smaller, with roughly 25% of the volume and weight of Skylab.

Many consider them to be merely smaller versions of Skylab — but direct comparison is difficult, for the Salyut programme is geared to developing multiple pieces of hardware which may be mechanically interlocked to perform tasks beyond the capability of a single Salyut vehicle. So far we have seen only one member of this family, namely Progress, which is a basic Soyuz vehicle fitted with automatic guidance and devoid of a life-support system; instead, the available space is used for fuel, food and water supplies, together with materials and equipment for experiments and data-gathering.

Transfer of fuel is a major step, for it allows the station to be manoeuvred and also to be periodically boosted in orbit to compensate for the inevitable decay due to drag from the

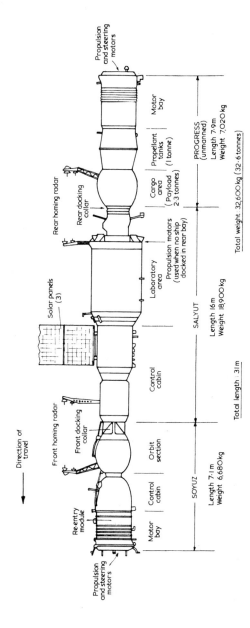

Direction of travel

Re-entry module

Propulsion and steering motors

Motor bay

Control cabin

Orbit section

Front homing radar

Front docking collar

SOYUZ
Length 7·1 m
Weight 6,680 kg

Solar panels (3)

Control cabin

Laboratory area

Rear homing radar

Rear docking collar

Propulsion motors (used when no ship docked in rear bay)

SALYUT
Length 16 m
Weight 18,900 kg

Total length : 31 m

Cargo area (Payload 2·3 tonnes)

Propellant tanks (1 tonne)

Motor bay

Propulsion and steering motors

PROGRESS (unmanned)
Length 7·9 m
Weight 7,020 kg

Total weight 32,600 kg (32·6 tonnes)

4 A complete Salyut 6 complex. The diagram shows two craft in dock — Soyuz at the front and Progress at the rear — orbiting Earth as a single unit. The Salyut normally is intended to orbit at about 350 km above the Earth's surface.

remaining minute traces of the atmosphere at that altitude. The engine of the 7-tonne Progress was used for this purpose on 6 February 1978. The automatic ferry was then filled with waste, undocked and deorbited and allowed to burn up in the atmosphere on re-entry. This operation was commenced on 7 February 1978 and Progress 1 burned up over the Pacific on 8 February. Novosti (Soviet Press Agency) remarked: 'This opens up the prospect of space stations being serviced from Earth by a reliable transport and *shuttle* service over a *period of years*. It also creates the possibility of building *complex engineering units in space.*' (My italics.) I think this is a significant statement for, to release the full potential of the Salyut system, at least three more units should make their appearance over the next few years. They are:

A Space Tug or *Kosmobuksir* (spacecrane): This might be employed in towing and manoeuvring various structures into place. This could be used for station assembly or for placing instruments into or recovering them from a higher orbit. Kosmobuksir is believed to be a development of Soyuz.

A Double-ended Docking Link: This might form a link assembly between (say) two Salyut 6 stations each fitted with two docking points. Initially the experimental link might be a straight tube which merely joins the two craft together to form a complete unit. Later versions might be curved or angled to allow the linking of six or eight Salyut 6 type stations to form a circular structure rather like a gigantic automobile tyre.

A True Shuttle: This item is already in existence, and has been reported as 'undergoing drop tests' from a TU 95 Bear during 1977-78 in similar manner to the glide tests performed on the US Shuttle Enterprise. The Soviet vehicle is called Kosmolyot (space yacht) and is approximately half the weight of the US machine. It is possible that instrumented test versions ('boiler-plate' models) have been flown in space in two mystery launches — Kosmos 881/882, which took place on 15 December, 1976, and Kosmos 997/998, launched in February 1978. It is likely that Kosmolyot will be used on military reconnaissance missions, but will use the Salyut as a Mother

Ship for extended mission work. In this rôle it could refuel and redeploy in a different orbit — and then return either to Earth or to the Salyut spacebase. (A more detailed comparison between Enterprise and Kosmolyot is made later in this chapter, but here we may note that it now seems certain that Soviet space science is developing along different lines to those of the West.) The Salyut 6 configuration may therefore emerge as the USSR standard manned module, with further modular units added until a complex capable of supporting (say) 18 to 20 cosmonauts has been created.

The Soviets have almost always followed what may be termed the modular method in developing complex equipment, and have used it in designing launching vehicles. At present two 'families' of launch vehicles appear to be used in manned flight work — Soyuz and Proton.

Soyuz is a member of the 'A' type launcher family. Designated as A2 by Western observers, it is a well tried system and was developed from a configuration used to launch Sputnik, Vostok, Voskhod and some of the early Luna vehicles. Depending on configuration and payload, the Soyuz vehicle launch weight is about 420-450 tonnes, and the total thrust of the combined first stage is approximately 510 tonnes. The overall performance of the Soyuz launcher is basically comparable with the US Titan 3c used to launch the two-man Gemini capsule (1964-66).

Proton is a member of the 'D' type launcher family, and the Salyut Proton has received the Western designation DI-H. Larger than the Soyuz, the Proton rocket employs the same strap-on booster principle for the first stage. This led to considerable development problems — the most likely cause of which was break-up of the vehicle at boost separation. These problems now appear to have been solved and Proton may be considered 'man-rated', although all Salyut launches to date have been unmanned.

Proton has been used also to launch the heavier Luna spacecraft, the Mars and Venus lander probes and the Zond lunar orbiting vehicles. A further developed version of

Proton will most likely be used for the early phases of Kosmolyot flights.

The take-off weight of the Proton is approximately 1,200 tonnes and, with a take-off thrust of some 1,500 tonnes, the vehicle is about halfway between the US Saturn 1B and Saturn V but grossly inferior in payload-lifting performance. Saturn 1B, with half the thrust (744 tonnes) placed Skylab at three and a half times the weight into a higher orbit than that normally used by Salyut. A Proton of 1,500 tonnes take-off thrust but a comparable *payload* performance to Saturn 1B *should* be capable of lifting nearly 160 tonnes of 'equivalent Salyut'!

These figures have been compared in some detail for they are the technical basis of the differing tactics and policies of eastern and western space programmes. It is difficult to understand why the Soviets have not exploited the benefits of rockets using liquid hydrogen, which, as explained earlier, offer a performance capability which is unmatched except by nuclear rockets. Unless Soviet designers accept the severe restrictions set by their present vehicles, a new and larger launch rocket with heavier payload capability must eventually appear.

The TT-5 G Class Superbooster
Rumours of such a launch vehicle began to circulate in the West in the late 1960s. Although it has never been formally admitted by the Soviets, the project involved the design of a rocket larger than the Saturn V. The launch weight has been estimated as 3,500 tonnes and the lift-off thrust as 5,000 tonnes. Payload has been estimated as 160 tonnes for a 'low orbit' (160km perigee).

The project encountered numerous setbacks. The prototype caught fire and exploded on the launch stand in early June 1969, and two other vehicles broke up after leaving the launch pad; at the end of 1972 all work appeared to have been suspended.

The US magazine *Aviation Week* for 8 May 1978 published a few details of future USSR experiments for lunar orbiting and landing vehicles. These will be much larger than the existing Luna series and imply a requirement for a much larger launch vehicle than the existing Proton.

One may speculate as to whether the Russians have successfully solved the problems associated with the TT-5 or whether they have high-energy upper stages almost ready for launch trials. Fitment of these to Proton could almost double its payload capability. Any success in such unmanned exploration could lead to a resumption of the Soviet manned lunar programme. We must wait and see.

The Shuttle

While the Saturn V launch vehicle was the critical link in the success of the Apollo lunar programme, and undoubtedly demonstrated the enormous lifting capability of a high-efficiency rocket, it did not impress the economists. The high cost of these operations was due mainly to the fact that the expensive launch vehicle was used only once. Most of it either disappeared into the Atlantic or was burned up in the atmosphere.

The costs are directly analogous to flying a Boeing 747 from New York to London or Paris — and then totally destroying it after the passengers have disembarked; another brand new 747 is then purchased and used for the return journey, and it in turn is completely scrapped on its arrival at New York. A passenger aircraft is made to operate economically by being designed for continuous recycling and reuse. This involves refuelling, replacement of consumed items and worn parts, and repair of breakdowns — all without compromising safety or reliability.

Not surprisingly, therefore, the NASA strategists decided to concentrate on *economic* justification of further manned exploration and exploitation of the space environment. The solution adopted by the designers was also a logical choice — an aerospacecraft. It would be a vehicle which could take off as a rocket, operate for up to 28 days in space without further aid, and, after a spacecraft atmospheric re-entry, fly through the upper atmosphere as a supersonic glider and finally touch down at an airstrip much as a conventional airliner does. The concept is not new — Dornberger studied an aircraft-lifted rocket-driven aerospacecraft design in the immediate post-war years; in the initial stages the NASA proposal was for an aircraft-lifted vehicle. Further heavy economic restraints, however, favoured a compromise design whereby the aircraft first stage was replaced by high-efficiency rockets which could be recovered and reused. The design finally hardened into what is now termed the 'Shuttle Programme' with the actual vehicles known as 'Enterprise'.

The concept represents a formidable step forward in manned space flight — and particularly as applied to astronomy. For the 18.3 x 4.6 metre cargo bay of the Enterprise is capable not only of taking 29.5 tonnes into orbit, but also of *returning 29.5 tonnes back to Earth*. Furthermore, the use of an 'aircraft' allows the handover of experimental packages for detailed analysis almost immediately after touchdown. This is not only good economics — it may be vital for certain experiments.

Each orbiter built is expected to make one hundred more missions, and this reuse is expected to reduce the cost of future manned space work by up to 90%.

Originally, there was one truly expendable item in the Shuttle concept, the external fuel tank (ET), but NASA is now studying various ways of using even the spent ETs. By adding solar electric panels, an airlock and a docking adaptor, the ET may be converted into a space platform able to support a crew of three for six months without resupply. Other ETs could be added as they become available, and studies indicate that they could be converted into factories manufacturing exotic materials difficult to fabricate on Earth. In this manner, a space station could be assembled with proven items in a short time and at low cost. An initial launch might be targeted for 1983-84.

New Scientist for 26 May 1977 published details of the possibilities of further US-USSR space programmes for the 1980s. There is already an agreement between NASA and the Soviet Academy of Science for talks on joint programmes.

Recognizing the different approaches that have been made by both sides, an attractive proposition that has been put forward for discussion is to use the Shuttle to ferry people and supplies to Salyut. By the 1980s the Soviet vehicle may have been enlarged along the lines indicated earlier. The Shuttle would be an efficient solution to the problem of economically shipping supplies and personnel to the space station.

No serious difficulties are foreseen in technical matters. A docking module would be used for linking the craft in similar manner to the Apollo-Soyuz joint mission of July 1975.

In some respects the proposed programme would favour the Soviets, who at present do not have a comparable vehicle to the Shuttle. Even Kosmolyot's cabin carries only a four-man crew, as opposed to seven for the Shuttle.

NASA should, in the proposed programme schedule, already have their own space station operational by the 1980s, although this may be delayed by Congress budget cuts. It is possible that NASA hopes that cooperation would result in the Soviets contributing to the budget and effecting further economies on the US side. Economics can be a powerful stimulant to enterprise.

Enterprise and Kosmolyot — A Comparison

Reference has been made throughout this chapter to the existence of a Soviet shuttle. Information on such a vehicle was received in London around 1970, when it was reported that the Russians were engaged in deep studies of the possible role of a spaceplane. At the International Astronautical Federation (IAF) Conference held in Amsterdam in 1974, Leonov confirmed that the Soviets were pursuing a manned shuttle programme. Further information continues to dribble through to Western observers. This has given details of 'drop-and-glide' tests carried out on a prototype. Combining this new data with the known performance of the most likely launch vehicle (Proton) enables a comparison to be made with the US Shuttle — see Plate 18.

This has been done in the two tables below; the first compares the main parameters of the vehicles, while the second deals with the launchers. Some artists have shown Kosmolyot being released from the back of a winged flyback booster, but my personal opinion is that this is most unlikely at present. I have, therefore, based the estimates of dimensions and weight on the

49

performance capabilities of the Proton launcher: this would be in accordance with the Soviet practice of using a proven system where possible. The calculated figures for Kosmolyot shown here are entirely in keeping with it being a smaller vehicle than Enterprise. It is assumed that structure densities are similar for both vehicles; slight deviations will not seriously affect the derived dimensions for Kosmolyot. The lower payload-to-weight ratio stems directly from the probable use of propellants with a lower efficiency than those used by Enterprise.

SHUTTLE COMPARISON

Item	Enterprise	Kosmolyot
Length	37.2 metres	25.5 metres
Height (to top of fin)	17.4 metres	10.0 metres (2 fins)
Wing Span	23.8 metres	16.5 metres
Weight	65 tonnes	20 tonnes*
Payload	29.5 tonnes	5 tonnes†
Crew and Passengers	7	4
Typical Orbit	300-500 km	300-500 km

*Based on the payload performance of Proton
†Based on the use of propellants other than liquid hydrogen/liquid oxygen

LAUNCH VEHICLE COMPARISON

Item	Enterprise	Kosmolyot
Lift off Weight	2000 tonnes	1200 tonnes
Lift off Thrust	3044 tonnes	1500 tonnes
Number of Stages	.2	4
Configuration	2 Strap-on Boost Motors	6 First-Stage Strap-on Motors
	1-3 Motor Central Stage	1 Central Stage
		2 Upper Stages
Propellants	1st Stage Solid Propellant	1st/2nd Stage LOX/KER*
	2nd Stage LOX/LIH†	Other Stages — Storable liquids

*LOX/KER = liquid oxygen and kerosene
†LOX/LIH = liquid oxygen and liquid hydrogen

5 A dramatic moment in the launch of a manned space vehicle, one of the Soyuz series. The scale of engineering required is shown by the relative size of the final check-out crew to the left of the rocket. (*Courtesy Novosti Press Agency*.) See page 45.

6 The second man on the Moon. 'Buzz' Aldrin, on 20 July 1969, descends the steps of the lunar module lander of the Apollo II mission. The photograph was taken by the first man on the Moon, Neil Armstrong. (*Courtesy NASA.*)

7 Lunar crater Daedalus (diameter 80km), which has a longitude of 179°E and a latitude of 5.5°S, thus being an ideal lunar farside observational site. (*Courtesy NASA.*) See page 59.

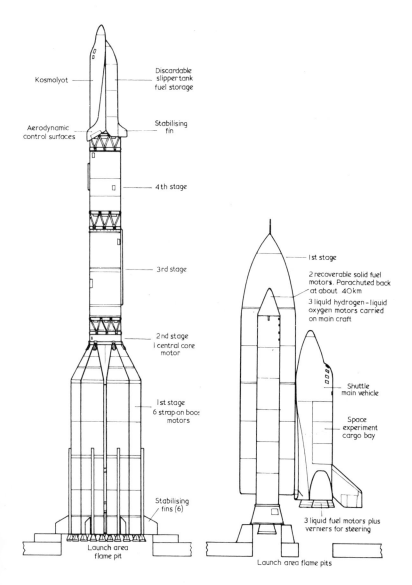

Kosmolyot

Discardable slipper tank fuel storage

Aerodynamic control surfaces

Stabilising fin

4th stage

3rd stage

2nd stage
l central core motor

1st stage
6 strap-on boost motors

Stabilising fins (6)

Launch area flame pit

1st stage

2 recoverable solid fuel motors. Parachuted back at about 40 km

3 liquid hydrogen-liquid oxygen motors carried on main craft

Shuttle main vehicle

Space experiment cargo bay

3 liquid fuel motors plus verniers for steering

Launch area flame pits

5 On the left is the probable launch configuration of the Soviet Kosmolyot as it would be if launched from Proton, the largest launching vehicle currently used by the USSR. From this configuration nothing is recoverable, but since the launcher is assembled from mass produced parts for the SS7 and SS9 ballistic missiles the cost of a launch is low. The U.S. Shuttle, whose launch configuration is shown on the right, reduces launch costs by reuse of existing items. The two drawings are to the same scale.

Enterprise will be the backbone of many future space exercises and this especially applies to astronomy. The USA is planning the assembly in space of large telescope structures — an almost impossible task without men and without the Shuttle as a ferry vehicle. Enterprise *is* the space equivalent of the Boeing 747 — it will be in use at least for the next two decades, and possibly beyond that.

The different characteristics for Kosmolyot may be indicative of a vehicle designed for missions of a different nature from those of Enterprise. In conjunction with Salyut such a vehicle could prove extremely flexible, with the ability to undertake large changes in altitude and orbit inclination.

In this and the previous chapter I have discussed ways and means of putting instruments and men into suitable orbits and accommodation above the atmosphere. But, about 4.5 billion years ago, Nature provided a site well suited for high-resolution astronomy 384,000km above the Earth's surface. It has firm foundations, and is free from atmospheric disturbance. It is called the Moon.

4 The Moon as an Observatory

Future Lunar Explorations

The Apollo manned lunar landing programme confirmed what had long been suspected. The Moon is utterly lifeless — and has been so ever since its formation and solidification about 4.5 billion years ago.

But bad news for the biologist is good tidings for the astronomer, and astronomers of the future should be able to establish useful observatories and bases in a uniquely suitable environment. Technology is about to provide the tools to do the job. Men have landed on the Moon and thereby demonstrated the feasibility of such a task, but the motivation has been purely political. Future Man will go to the Moon to explore and exploit, for the Earth's only satellite represents an enormous store of mineral wealth. And it is on this 'piggyback' of economic necessity that suitable low-cost transport vehicles will be developed. Once this breakthrough is established, astronomical observatories will be built on sites which have been selected much earlier in the original phase of exploration.

Such a scenario runs contrary to popular science fiction, but I personally doubt if a Moon base will ever be built for purely scientific reasons. Man's history has shown a consistently repeated pattern of development which may be summarized as: 'Explore, exploit, and create "wealth"; then use this wealth to support science and technology in order to solve the problem of the day; and thus assist further exploration and exploitation. . . .'

The mainstream of astronomical development has certainly followed this course from Stonehenge to Greenwich. The former seems to have been built to forecast eclipses, equinoxes and solstices; essential for the purposes of creating an agricultural calendar and forecasting important events with a religious significance. Greenwich Royal Observatory was specifically commissioned to prepare more accurate charts of the positions of the Moon and stars, not for scientific research, but for better navigation — making for faster and more efficient trading and hence greater creation of wealth.

I would suggest that lunar observatories will be founded in a similar manner. The major manned expeditions to the Moon will be exploratory and for the purpose of charting and surveying the topography and possible mineral concentrations. These manned expeditions will follow thorough reconnaissance by unmanned vehicles which will orbit the Moon and take very high resolution images in infrared and visible light. Selected areas will be drilled and the cores returned to Earth for analysis. Those samples with the most interesting results will be further drilled and examined and then designated as areas for manned exploration. Already, from the Apollo experiments, we expect areas rich in silicon, aluminium, titanium, manganese, magnesium and iron. All of these materials are useful materials for structural purposes.

In his recent book, *The High Frontier: Human Colonies in Space*, G. K. O'Neill describes the recovery of such material from the lunar surface for use in constructing very large cylindrical structures (8km in diameter, 32km long). O'Neill's analyses show that use of such lunar processed material is far more economical than use of material ferried up from Earth.

It is on *this* scenario that I suggest Moon observatories will be built. I use the plural, for at least two would be required. The first would be built on the side facing Earth simply because that is where the initial material-recovery plant would be sited. The second would be sited on the far side of the Moon where it would very nearly have a true 'dark night'. This would have very definite 'seeing' advantages in several wavebands from X-ray through uv, visible, ir down to radio.

The Lunar Observatory

The Moon offers an observatory site with several major advantages, among which are the following: it is without a measurable atmosphere; there is no 'weather' in the form of wind or rain; the Moon is large enough to provide a stable and firm foundation; the lunar crust is thicker and much more rigid than that of Earth and seismic activity, in the form of Moonquakes, is at a much lower level than on Earth; there is little volcanic activity.

All of the above items allow the construction of buildings which have to endure extremes of temperature, but little or no weathering or other natural 'interference'. They will not be rusted or corroded, or be scoured by sand. Lunar structures may need to be stressed to withstand pressurization to (say) 0.5 bars, half of Earth's atmospheric pressure at sea-level: this corresponds to the atmospheric pressure on Earth at 6,000 metres. Further stresses are very low, for the lunar gravitational field has a strength of only 15% of that of the Earth's; i.e., a telescope weighing 100 tonnes on Earth would weigh only 15 tonnes on the Moon. (However, it would still possess an inertial mass of 100 tonnes and would require drive gear and controls to suit.)

Finally, the low seismic activity would make for a stable site which could be a useful feature in astrometric work. This consists of taking extremely precise measurements of star positions; amassed over a long period, these measurements can reveal the presence of small invisible companions to nearby stars.

The Moon offers other attractions as an observatory site, and these additional advantages lie in the peculiarities of the lunar orbit. Our satellite has captured rotation — that is, it rotates on its own axis in the same time taken to complete one orbit of the Earth; this period is 27 days 7 hrs 43 mins 11.5 sec, or 27.3215 days. An *average* lunar day and night would therefore be approximately 13.66 Earth days, and *theoretically* a lunar astronomer could thus train his instruments on an object for a period forty times longer than could his Earth-bound counterpart.

I stress the word 'theoretically' for such a statement is based on the 'averaged' lunar day and implies that a Lunar Observatory may be placed anywhere on the Moon and enjoy the above advantages. Not so. The site for any Lunar Observatories must and will be selected with all the care and deliberation lavished on the present-day terrestrial sites.

Choosing a Lunar Site

The movement of the Moon through the heavens is extremely complex and is repeated in precise detail every 18 years, 10 days, 7 hrs, 42 mins — the Chaldean Saros. Therefore an account of the sky coverage available for observation at any potential site must include this full cycle.

Some of the site requirements will be in conflict. For example, it is obviously desirable to be in communication with Earth — and yet the siting of the observatory on the side *facing* the Earth poses a far worse stray light problem than full Moon does to terrestrial astronomers: full Earth to a lunar astronomer would be about 70 times as bright (an increase of 4.6 magnitudes). This is due not only to the larger reflecting area but also to the higher reflectance of Earth clouds; the Moon reflects only 7% of the incident sunlight while the Earth reflects 34%.

However, safety of personnel would be paramount, and it is almost certain that the first observatory will be sited facing Earth and will carry out tasks which are not embarrassed by stray light.

Lunar astronomers will pay a great deal of attention to possible polar observatories. Since it would not be possible to observe all of the sky from one polar site, both North and South lunar poles would be used. Two observatories so situated would offer several advantages over an equatorial or more conventional system. By suitable choice of crater it may be possible to observe in lengthy — although not the theoretically ideal — periods of total darkness. This is because the Moon's orbit around Earth is inclined 5° 8' to the ecliptic. At the lunar poles, therefore, the Sun will be low down on the horizon during the lunar day, thus reducing temperature extremes. During the lunar night the Earth will also be low on the horizon and, due to an effect known as libration in latitude, there will be periods when both Sun and Earth are below the horizon. The libration referred to is due to the Moon's equatorial plane being displaced 6.5° to the plane of the lunar orbit, which allows the north and south poles to be alternately presented to Earth during the course of one complete orbit.

Depending on several factors — e.g., if the chosen site is a crater, the height of the walls — the lunar astronomer would have two total-darkness periods per month. Each period would probably last not more than three or four days. After that time, direct and crater-wall-reflected light would gradually increase until either Sun or Earth was again above the horizon.

Continuous direct radio communication with Earth is not possible from any one station, and in order to maintain such contact a chain of satellites would be required.

I think that these polar stations will be a logical step towards the final goal, which is the farside observatory where the Earth is not visible at any time and consequently astronomers would have the advantage of the long lunar night of almost 14.5 days. The most logical way of doing this is to go 'over the poles' a further distance which corresponds to just over 6.5° of lunar latitude. If we chose 7°, to be on the safe side, this would correspond to a distance of 212.5km. Thus, by pushing on this extra distance, we gain all the assets of a true farside observatory, and I think that this is exactly how the initial farside stations will be built. It is the logical 'advance, consolidate, exploit' scheme mentioned earlier in this chapter.

Farside stations present risks as well as advantages, for at no time will they ever be in direct contact with Earth. All communication will be *via* satellite or ground station relay. Furthermore, on an equatorial farside station there would be enormous changes in temperature, ranging from 420 K during mid lunar day to 70 K just before lunar dawn. Equipment and living quarters would have to cope with these extremes. This is another point in favour of polar region farside observatories. A possible farside crater is Daedalus (see Plate 7).

A Lunar Radio Observatory

It is possible, however, that radio observatories will move further 'over the poles' than optical sites. This is because the Earth (and to a lesser extent the nearside of the Moon) will be a source of manmade radio noise or 'static interference' which could wreak havoc on delicate radio measurements. Although optical instruments would not be affected, it is possible that radio waves from Earth could be bent or diffracted by the dielectric properties of moonrock and thus cause interference even when Earth was out of sight below the lunar horizon. The optical analogy of this effect can be seen on Earth when the 'setting' Sun is still above the horizon: although calculation will show that the Sun is already *below* the horizon, the Earth's atmosphere is bending the light rays such as to make it still visible to the observer.

In the case of the lunar radiotelescope, the cure would be merely to shift the installation further 'over the pole' and down onto the farside, another 100km from the optical site. The low gravity field and extreme stability of the Moon's surface allow the construction of enormous reflectors, or 'dishes' and the electrical matching of several such surfaces to act as one huge collector.

The largest dish that can be freely mounted and supported on Earth is 100m in diameter. Attempts to build larger items have so far failed simply because the materials used are not strong enough. Ingenious design has to date partly offset the problem, for before about 1968 the maximum attainable diameter was a mere 75m.

On the Moon, an object experiences only 0.167 of its Earth weight, so, assuming that a dish has a weight proportional to its *area* (a good approximation for thin objects), we can build reflectors with six times the area possible on Earth. Since area is proportional to the square of the diameter, we can build reflectors 2.45 times the diameter of terrestrial items using the same material. Our 100m Earth dish becomes a 245m Moon dish. Performance or sensitivity is proportional to area, so the Moon antenna dish will have six times the sensitivity of the Earth unit.

A radio signal is reduced in intensity in proportion to the square of the range travelled. A radio wave is therefore reduced to 0.167 of its power when it has travelled 2.45 times the range. In other words, the range of a radiotelescope is directly proportional to its antenna reflector diameter. If our 100m Earth telescope could listen for radio beacon signals from other stars out to a range of 100 parsecs, the Moon telescope could reach out 245 parsecs and thus cover fifteen times as many stars.

Possible Experiments in Lunar Astronomy

By the simple expedient of bouncing a laser beam off the Moon's surface and measuring the time delay between the outgoing pulse and its return, we have been able to determine the Earth-Moon orbital distance with considerable accuracy. This work was part of the ALSEP Apollo experimental package. As a sideline, this particular experiment showed that the surfaces of both Earth and Moon 'breathed' or bulged towards each other at perigee.

This allows us to do trigonometrical surveying in exactly the same way as is done on Earth when we are laying out roads or railways. The simplest system would be to use a lunar-based optical telescope in conjunction with one on Earth — say the 5m reflector at Palomar. Theoretically we could accurately measure the distance to an object — say a star — by simply pointing both telescopes at it, measuring the sighting angles and using our laser-measured Earth-Moon distance as a baseline.

In practice this would not be a good experiment for we have already determined the distances to the nearest stars by using twice the distance from the Earth to the Sun as a baseline. This is nearly 750 times that of the Earth to the Moon and, having been accurately surveyed, has been used as a baseline for nearly 150 years!

A better experiment would be to use the two telescopes as an interferometer. This system was first used by Michelson on the 2.5m Mount Wilson telescope. Michelson mounted two mirrors 20ft (approximately 6m) apart and arranged them each to reflect the light of the star being studied into the telescope aperture. The starlight is seen as a series of interference fringes and the distance separating the two mirrors is adjusted until this interference pattern just disappears. It can be shown from simple optical theory that the angular diameter (d) of the observed star is given by

$$d = \frac{1.22\ \lambda}{D}$$ radians, where λ is the wavelength of

the light and D is the separation distance of the mirrors.

In Michelson's work the wavelength used was 0.57 μm (i.e., 0.57 x 10^{-6}m) and, with the above separation, he was able to resolve the diameter of Arcturus as approximately 0.02 seconds of arc. He later attempted a system with the mirrors separated by 50ft (15.25m) but ran into severe practical difficulties, mainly due to the telescope flexing under the weight of the arm supporting the mirrors.

Theoretically it would be possible to construct a super-interferometer working at optical wavelengths and using a laser beam to 'tie' the telescope mirrors on Earth and Moon. In practice this would be impossible, for scintillation of the Earth's atmosphere and various other errors would destroy the interference pattern.

Interferometers can be made to function at other wavelengths, and because of the long baseline available between Earth and Moon we can choose a wavelength longer than Michelson was able to use and yet obtain a much higher resolution. For example, if we used a pair of radiotelescopes and operated at a wavelength of 1m, then, because we have a baseline 384 *million* metres long, we should be able to resolve

$$\frac{1.22 \times 1}{384 \times 10^6} \text{ radians,}$$

which is approximately 1.14×10^{-5} seconds of arc — almost 90,000 times better in resolution capability than an optical device. This degree of resolution would allow the detailed mapping at radio wavelengths of actual surface features on stars as well as resolution of the star disk itself.

However, all is not lost on the optical side, for the high stability of the lunar surface would allow the accurate construction and mounting of what is termed an *intensity* interferometer. In this apparatus the optical signals are mixed electronically. A system of this type is operating at Narrabri some 480km north of Sydney, Australia. The two 'mirrors' are very simple and do not need to be finished with the precision of a telescope. On the Moon they could be about 26m in diameter and separated by much greater distances than are their Earth counterparts. The increased light-collecting capability and resolution would make it possible accurately to resolve dwarf stars similar to or smaller than our Sun.

Lunar Pollution

The Moon may not have a very long life as an astronomical base, for the industries which support the research will also generate an atmosphere — and to such an extent that eventually telescopic seeing will be as bad as it is on Earth. The atmosphere may not be

Mineral	Formula	Percentage	Refined Material	Uses for Refined Material
silicon dioxide (quartz)	SiO_2	40.4	silicon	Integrated circuits for microcomputers. Silicon solar cells. Special steel.
iron oxide ('black rust')	Fe_3O_4	19.3	iron	Wide application, principally stressed structures and electric generators and motors.
calcium oxide (quicklime)	CaO	11.0	calcium	Lightweight structural alloys. Electrical batteries.
titanium dioxide (ilmenite)	TiO_2	10.9	titanium	Very highly stressed lightweight structures. Implants in bone and tissue surgery. Artificial limbs.
aluminium trioxide (alumina, sapphire, ruby)	AlO_3	9.4	aluminium	Lightweight alloys. Electric wiring and cables. Coating optical mirrors.
magnesium oxide (magnesia lime)	MgO	7.2	magnesium	Lightweight alloys. Electrodes for electrical batteries.

as dense as that of Earth, but 'clouds' will be made up of particles of dust rather than of moisture. This turn of events will be inevitable if we consider the Moon as a source of raw minerals.

Analysis of the rocks returned by the Apollo missions has shown that they are composed mostly of the oxides of minerals which are very useful to Man. They may be used in raw form as building material or refined to separate the oxygen. The table above shows the basic minerals in a typical sample of moonrock, the refined material, and uses for the refined products.

All of these materials are *very* rich in oxygen; for example, in quartz 53% of the total mass is oxygen. The rest of the materials have different amounts of the gas in their makeup, but it can be shown that oxygen accounts for just over 42% of the sample used for the table. If we mined 100 tonnes of similar rock we could theoretically recover 42 tonnes of oxygen!

It seems ironical that the Apollo astronauts were compelled to wear clumsy spacesuits and breathing apparatus — and yet they were standing on a surface whose major constituent was the gas that kept them alive.

Detailed studies and experiments have been done on moonrock (and similar Earth materials) to see if the purified materials listed above can be recovered. The studies centred on cooking the rocks at dull red heat (750 K) in a solar furnace and in an atmosphere of hydrogen. This yielded superheated steam which could be condensed to water and then split up into hydrogen and oxygen by electrolysis. (On the Moon, the power for this may either be obtained from solar cells or from turbogenerators driven by the superheated steam — there will be much scope for ingenuity!) In the Moon mining and metal-recovery installation, the hydrogen would be recycled and the oxygen initially used to support the human population. Eventually, however, there would be a surplus. This would be vented — possibly onto the Moon's surface or into a deep valley. There it would augment the minute traces of heavy gases (argon, krypton, xenon) which may lurk in deep clefts; these gases are believed to be the residues of radioactive decay, meteoritic impact and low-level volcanic activity, and are harmless. If venting were at a greater rate than the Moon's natural losses then an atmosphere of oxygen plus trace gases would form.

It is possible to obtain a very rough estimate of when this is likely to happen. The world's consumption of metals approximately doubles every 35 years — a trend which can be expected to continue as Third World countries reach Western standards of consumption. And if O'Neill's 'Habitat' cylinder programme is undertaken, then the materials will be recovered from the Moon. (According to Professor Brian O'Leary of Princeton, lunar material for these can theoretically be obtained at a cost of $100 per tonne compared with $1 million per tonne if obtained from Earth.)

If, then, a lunar atmosphere as dense as that of Mars be considered as a limit, then the mass of oxygen required will possibly have accumulated by the middle of the twenty-fourth century. This is an almost instantaneously short period in geological (or, rather, selenological) terms. It is therefore unlikely that the Moon will lose this manmade blanket at a similar rate and probably the waste oxygen will accumulate, together with other unwanted non-toxic gaseous by-products.

The effects on lunar astronomy would be disastrous! Long before the atmosphere had built up to the assumed density it would support superfine dust and disperse it on a global scale through the agency of very high-speed winds. This would coat optical mirrors and find its way into bearings and seals, and of course ruin the 'seeing' with atmospheric turbulence and dust.

This scenario presupposes that hydrogen will be available in sufficient low-cost quantity to reduce the lunar rocks to their constituent metals. Initially this hydrogen would have to be ferried up from Earth, but a high degree of hope is centred on finding water on the Moon. It is an essential ingredient of lunar economics.

Although not very widely reported, the orange soil discovered by the Apollo 17 crew has been analyzed: it derives its colour from hydrated iron oxide — better known as rust. In particular, the minerals goethite and lepidocrocite ($FeO.OH$) have been identified. The report further stated that the sample represented an *average* water content for lunar rock of 150 to 170 parts per million (0.015% to 0.075%). If this is correct then the Moon might have a total water content between 1×10^{16} and 5×10^{16} tonnes. This would be more than adequate for reducing the lunar raw materials, even when allowance is made for being able to recover only (say) 1% of the material available. Since the hydrogen is recycled, it is required only to make up the inevitable losses. The fully confirmed presence of limonite on the Moon would make the production of finished metal from local raw materials practicable — and so by AD 2350 useful lunar astronomy would cease.

However, it is possible that the production of an atmosphere with the density postulated would be self-defeating and positively hinder the mining operations. O'Neill proposes to use an electromagnetic linear accelerator to project 'slugs' of raw or refined material from the lunar surface to where they are required. Driven by solar power, the linear accelerator has been assessed as viable, efficient and economic, hence the low transport costs quoted by O'Leary.

The presence of an atmosphere would upset the efficiency on two counts: one, the solar power cells (of large area) would become coated with fine dust, thus reducing output; and, two, the slug could become intensely heated travelling through an atmosphere at the lunar escape velocity of 2.38 km per second. And if that atmosphere were almost pure oxygen, then slugs of magnesium, calcium and aluminium would burn with spectacular brilliance! Even if this were overcome by the use of a refractory coating of (say) rock slag, there would still be the extra energy required to overcome frictional atmospheric drag.

To sum up, I consider that lunar observatories *will* be built and *will* carry out valuable research over a wide range of wavelengths. They will be largely financed as 'piggyback' operations to a mainstream of lunar material mining and recovery. But this industry, if allowed to develop in an uncontrolled manner, could put an end to lunar astronomy. If, however, the industry is allowed to expand and grow in a *controlled* manner, both astronomy and mining should proceed side by side.

Diamonds in the Sky!

Hydrogen has been quoted as the reducing agent in these metal-winning processes, but carbon may also be used to reduce ore. No traces of this material have yet been found in lunar rocks, but on 3 November 1958 the Soviet astronomer Nikolai A. Kozyrev, working with a 1.25m reflector telescope at the Crimean Astrophysical Centre, was examining the crater floor of Alphonsus and observed a reddish patch near the central peak. Fortunately he was able to photograph the spectrum of the glow. The plates showed that a hot gaseous carbon compound was being emitted from the area of the peak. Lunar rocks are oxygen rich, so the most likely candidates were carbon dioxide and (possibly) carbon monoxide. Kozyrev estimated the temperature of the emission as about 1,400 K and its total duration about 30 minutes. If there *is* carbon on the Moon, then this will save one more raw ingredient being ferried up from Earth — at $1 million per tonne its cost would be in the same ball park as that of diamond, the exotic crystalline form of carbon.

And it is conceivable that lunar carbon *will* be found in meteor craters in small quantity in the form of diamond, for some of the impacting meteorites will have been carbonaceous chondrites. Carbon forms about 30% of their total mass. If such a meteorite smashed into the lunar surface, then under the intense heat and pressure generated a quantity of this carbon would be transformed into diamond. Much of it would be lost when the impacted surface 'rebounded' and so threw the material out, but a fraction of it might be retained in the vicinity of the central peak. Only a detailed exploration can prove this, but it may be significant that Kozyrev observed hot carbonaceous gas from the area around the peak of Alphonsus, and there may be several craters with similar characteristics.

ALSEP — The End of the Beginning

While the popular news media were mainly concerned with the more trivial aspects of the Apollo lunar landings, a more important experiment was in progress throughout the later phases of the programme. Called ALSEP (an acronym for

Apollo Lunar Surface Experimental Package), the experiments were designed to measure — and transmit to Earth — data on temperatures, meteoritic impacts, seismic shocks (Moonquakes), etc. A laser reflector suitably oriented allowed signals to be transmitted from Earth and received back with sufficient intensity to allow the Earth-Moon distance to be measured with extreme accuracy.

The prototype was installed by Apollo 11 and lasted 70 days, but the equipment installed by Apollos 12, 14, 15, 16 and 17 lasted a grand total of 29 years. The Apollo 12 ALSEP was 8 years old and still 100% functioning when the whole system was closed down, primarily for economy, on 1 October 1977. During their lifetimes these five equipments received a total of 153,000 commands and transmitted 10 million 'bits' of data.

The experimental results show that the Moon is a *very* stable platform for astronomical instruments. The seismic energy released in Moonquakes is about 10^{15} ergs per year, whereas over the same period the Earthquake energy release is 5×10^{24} ergs — *5 billion* times more. (10^{15} ergs is roughly equivalent to 200 tonnes of TNT; 10^{24} ergs corresponds to 1 million megatonnes). The tremors experienced by the Moon also seem to occur at regular periods. These correspond to perigee (closest approach to Earth) and may be likened to the Moon 'breathing' or being stretched by the gravitational pull of the Earth.

The apparatus also effectively demonstrated that meteoritic impacts were about one hundred times less than expected.

ALSEP was switched off because it had completed its task. It demonstrated that the type of equipment required for observatories would find a stable environment on the Moon. We have seen the 'end of the beginning'; the next phase will surely follow.

8 & 9 Mariner flybys of Mercury showed that the planet is, in many ways, like the Moon, but there are a number of important differences. *Above* is shown the Caloris basin, the result of the impact of a planetoidal body of mass and velocity almost sufficient to shatter the planet, according to many planetary geologists. In the photograph *below* a lobate scarp is indicated; such scarps, which would seem to indicate that there was at one point, after the formation of nearly all of Mercury's other surface features, a global shrinking, with a resultant wrinkling of the crust. (*Courtesy Jet Propulsion Laboratory, California Institute of Technology.*)

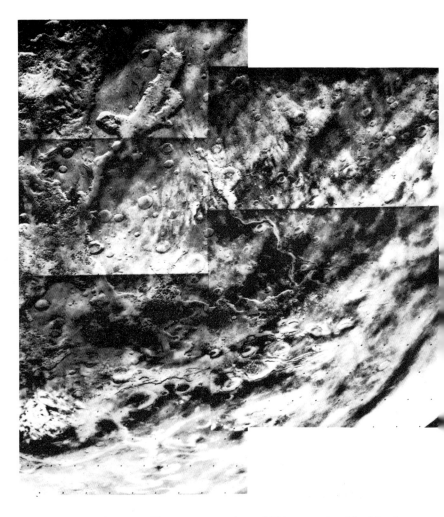

10 From a distance of just over 30,000km, a Viking mosaic of the Martian surface. Gangis Chasma, at lower right, is a canyon that leads into the stupendous Valles Marineris complex, some of which is visible along the right edge of the mosaic. The white patch near Gangis is thought to be ice fog close to the surface; the white patch in the top right-hand corner is an ice cloud moving toward the lower left at about 150kph. (*Courtesy NASA.*) See page 91.

5 Roentgen Revisited

In Chapter 1 I mentioned the discovery of X-rays by Wilhelm Roentgen in 1895. In just over 80 years since his discovery, this short-wavelength radiation has not only worked miracles in the medical world (e.g., the X-ray whole body scanner) but has enabled astronomers to make discoveries which were (and continue to be) totally unexpected.

We now associate X-rays with short wavelengths and tremendous energies: not only are X-rays about 1,000 times shorter in wavelength than is violet light, they have 'photons' at least 1,000 times more energetic. A photon is a particle which has no mass, but moves at the speed of light and vibrates at a characteristic frequency which corresponds to its wavelength in the electromagnetic spectrum. The greater the frequency with which it vibrates, the more energy it possesses. Each photon therefore corresponds to a definite 'packet' or quantum of energy.

It is in this context of possessing large inherent energy levels, and therefore being the signature of energetic events, that X-rays have assumed such importance in the astronomical field. Nature generates X-rays very efficiently and in copious quantities, and it is the production of large quantities of extremely energetic photons that astronomers are presently endeavouring to explain — albeit with a good measure of success.

We cannot at present duplicate the conditions under which Nature generates X-rays so efficiently. This is a blessing, for this form of radiation is extremely destructive to human tissue. We can stand a lifetime of sunbathing in violet and ultraviolet light, as filtered by our atmosphere, but if we were subjected to the same intensity of bare skin bombardment by X-rays then we would last about a month before dying of extreme sunburn and other biological complications. There is a major safeguard against this happening: the atmosphere is entirely opaque to X-rays. Moreover, the level of solar X-ray emission is well below that of visible light; if it were otherwise it would mean that the surface temperature of the Sun was very much higher, which in turn would mean that the Earth would be not only uninhabited but uninhabitable.

But, as we shall see, there are certain classes of stars and other objects which *are* extremely hot and do emit copious quantities of 'Roentgen's Rays'. Only now are we beginning to realize that his discovery is causing as much of a revolution (and revelation) in astronomy as did Newton's discovery of the visible light spectrum — and it is possible that new and further X-ray discoveries will keep us occupied for the next century!

'In the Beginning'

X-ray astronomy got off to an unexpected surprise start on 18 June 1962 when an Aerobee rocket carrying an X-ray detector and collimator was launched in the hope of picking up X-rays expected to arise from the bombardment of the Moon's surface by solar radiation. No lunar X-rays were found during the brief period the rocket was aloft, but a brilliantly powerful X-ray source was found in the constellation of Scorpio. In accordance with normal astronomical practice this was named Sco X-1. The Aerobee and similar small 'sounding' rockets continued to be the main source of information for the next eight years. Balloons provided a longer-duration platform, but even these were not sufficient for they rarely allowed flights to last longer than 24 hours and there was still the problem of residual atmospheric shielding since the system was not truly in space. Background noise produced by unwanted radiation showers from cosmic particles impacting the atoms of the thin traces of the atmosphere also hindered accurate X-ray measurement of the fainter objects. These early instruments, crude though they were, allowed the design, development and production of efficient detectors and collimators such as to allow the precision location of X-ray objects and subsequent division of the radiation into suitable levels.

In December 1970 the X-ray satellite Uhuru (Freedom) carried out the first three sky surveys at X-ray wavelengths. The last of these (the 4U catalogue) defined 160 new sources. Since then other satellites have been launched, the most successful being Ariel V, launched in 1974 and still pouring out data five years afterwards. Other satellites are planned or have recently been launched.

X-ray Generators — Binary Systems

Now, after nearly twenty years of study, we have explanations as to how X-rays are generated. After the 1962 launch, further measurements from instruments lofted into space for short periods showed that Sco X-1 was a variable X-ray source and of small angular diameter. Armed with position measurements, astronomers hunted for an optical counterpart and found a faint but *very* hot star which emitted copious radiation in violet and ultra-violet light. More important, the optical radiation varied synchronously with the X-rays.

This was again a lucky discovery, for Sco X-1 is one of the very few X-ray sources that is also optically visible. In the majority of cases, the companion star (usually a blue giant or supergiant) can be seen but not the actual source itself — for the latter is usually a neutron star or black hole.

Sco X-1 has a light output variation which is very similar to certain types of novae, particularly 'old novae' where the outer shell is absent. The hot central core is believed to be that of a white dwarf with a skin or top surface that is still violently active although the actual nova outburst may have occurred some decades (or even centuries) earlier.

At about the same time that Sco X-1 was discovered, theoretical and practical work on several novae showed that they were *binary* systems with one of the partners a white dwarf and the other a larger star. *All* of these binary systems are characterized by a small separation between the two components. They are termed 'close' binaries and in extreme cases the separation may be as little as a few million kilometres. Under these circumstances the intense gravitational field of the white dwarf pulls and distorts the atmosphere of the larger star to such an extent that it spills over and cascades onto the superhot surface of the dwarf. Here it is compressed and heated to a sufficiently high temperature for it to undergo thermo-nuclear reactions. Hence the blaze of radiation that forms a nova.

The strong resemblance of Sco X-1 to the behaviour of a nova suggested that it was a binary star system. There is a slight periodicity in the X-ray and optical output of Sco X-1 which suggests that this is the case, but the fluctuation is small and we may be looking 'down on' the system and therefore not seeing a true eclipse.

A specialized form of X-ray binary is where one or more of the partners is a black hole. In the case of Cygnus X-1 the black-hole concept seems to have been proved 'beyond reasonable doubt'. The visible companion is a blue giant of about 30 solar masses (as required by theory) and the X-ray source is of 8 solar masses. Other candidates require further study (see Chapter 11). The generation of X-rays from these objects is on an enormous scale; for example, the output of Sco X-1 at X-ray wavelengths is 10^5 (100,000) times that of the Sun's total output. It has been virtually constant at this level (with minor cyclic traces) since first observed.

The energy sources available in a binary star system are in the spin of the stars, their orbital periods, and the gravitational fields between them. In the close binary systems, the latter source dominates *everything* and is responsible for the X-ray production by acceleration of matter from one star (generally the larger) to the other and the conversion of the acquired momentum to heat. This raises the temperature to nuclear reaction levels. The flare-up is estimated to involve the loss of about 1/10,000 of a solar mass, but in the case of Sco X-1 the process is continuous. This constant outpouring of X-radiation has been observed in a few other cases, and it is plain that this phase of a close binary is short-lived: it is doubtful if such an output is maintained for longer than 10^4-10^5 years. The mass radiated away as energy allows the stars to drift apart — i.e., the binary opens out — and consequently the X-ray production becomes intermittent. The tap is only turned on when the stars undergo their closest approach, and stops again when they move apart. This type of system is more frequently encountered, and it is tempting to speculate and suggest that white dwarfs may become neutron stars or black holes through being 'stoked up' by material acquired from the companion.

Crab *à la* Roentgen

A sophisticated rocket experiment was performed in July 1964 when an Aerobee carried an X-ray telescope above the Earth's atmosphere for a long enough period to allow the apparatus to record an occultation of the Crab Nebula by the Moon. As the nebula was covered by the lunar disk the variation of X-radiation showed that the source of emission had a large angular diameter rather than being a point.

The nebula was found to be radiating about 10^{30} watts in X-rays, and, although Sco X-1 was the first discovered X-ray source, the Crab was the first source to be positively linked and identified with an optical object. The power level is about 10,000 times the total power output of the Sun.

The output from the Crab was initially thought to be continuous, not cyclic as with a binary X-ray source, and it therefore must have a different method of producing X-rays. In 1953 Iosef Shklovskii had proposed that synchrotron radiation was responsible for the weird blue light emitted by one of the stars at the centre of the nebula. The same explanation has now been invoked for the production of X-rays. Radio pulsars and gamma-radiation are also most simply and easily explained by this mechanism, which involves electrons spiralling along magnetic fields. If the electrons are moving and spiralling such that their total velocity is at or near the speed of light, then relativistic effects set in. The electrons cannot move any faster, but if an accelerating force is still effective then somehow the spiralling electrons *have* to shed the surplus energy. This they do by producing electromagnetic radiation.

More precisely, in quantum terms the electrons emit photons, and, since quantum mechanics allows statistical degrees of freedom, the radiation may be emitted at several wavelengths. A given amount of energy may be shed by emitting a few photons with enormous energy per photon, in which case we have gamma-rays. Emission of a larger number of photons at a lower energy level allows X-rays, still more gives visible light and finally a large number of low-energy photons results in radio emission. All of these states are possible at any time, and the result is a continuous spectrum.

Following the discovery of the pulsar, efforts were made to detect X-ray pulses and in 1969 H. Friedman and his team of the US Naval Research Laboratory confirmed that the Crab was indeed producing X-rays at pulsed intervals. These pulses have been accurately plotted and their outline or waveshape with respect to time is almost identical to the radio pulse shape. In fact, from radio frequencies of 300 MHz (1 metre) to X-rays of 3×10^{20} Hz (10^{-12} metres) the waveforms are virtually duplicated in time and shape. Such uniformity is strongly indicative of a common mechanism for generation — convincing supporting evidence for Shklovskii's synchrotron theory.

The test of synchrotron radiation is the measurement of a high degree of polarization. Since this can be observed with the radiation from the Crab, it seems virtually certain that synchrotron radiation is responsible for the X-rays.

The total X-ray output from the Crab is (as stated) about 10^{30} watts. It is made up of two components, a steady output of 10^{13} watts from the nebula cloud and a pulsating component of 10^{17} watts from the central neutron star fragment.

The Crab continues to be one of the most interesting astrophysical laboratories in the heavens and no doubt will be the subject of many future studies.

The X-ray Skies are Hot!

X-ray objects come in a variety of forms, and among these are the remnants of supernovae other than the Crab. The remnants are in two parts, an outer ring or loop of filaments which are the vestiges of the outer shell of the exploded star, and a compact source which represents the inner — and probably compressed — core. A good example of such a set of remnants is the Cygnus

Loop. The loop itself is a large 'sickle' of glowing gas almost 32 parsecs in diameter. In the centre of this sickle-shaped loop is a small almost pointlike source of X-rays which could possibly be the crushed core of the star. We cannot be certain since there is no pulsar (radio, optical or X-ray) to guide us. The loop itself is also a source of X-rays. The radiation generation by the loop is different to that of the central core, and is believed to be due to the outgoing shell colliding with the atoms of interstellar gas and heating it to several million degrees. The central element is a synchrotron radiation generator.

The Cygnus Loop is believed to be about 50,000 years old, and since it is fairly close (600 parsecs) the original outburst must have been very impressive, for it would be about four times as bright as the great supernova of AD 1006 and could have reached a visual magnitude of −11.0. This corresponds to Full Moon, but *this* brilliant light would have appeared to come from a point source and would have scintillated or twinkled unbearably, making it impossible to look at it for long without being dazzled.

In addition to the Cygnus Loop and the Crab, the supernova remnants of the Vela, Cassiopeia A, Tycho's supernova of 1572, and the great supernova of 1006, are sources of X-ray emission. Other sources that have been noted are Puppis A and IC 443. Possibly a ninth supernova remnant found in Cygnus in 1977 may be included on the list of X-ray remnants, although further confirmatory work is required.

All of the objects discussed so far are within our own Galaxy, but it is now known that immense clouds of intensely hot gas exist between galaxies. The intensity of the X-ray emission gives some idea of the mass of intergalactic gas. It seems that this 'invisible X-ray Universe' has a mass which is *at least* equivalent to the mass of the visible Universe.

If this is verified then X-ray observations will have resolved the paradox of the 'missing mass', thereby allowing a 'closed' Universe. At present the Universe is thought to be 'open'; i.e., it is flying apart from the initial push of the Big Bang and according to some authorities it will go on forever. This is because there is not enough material to allow gravitational perturbation to slow the expansion down; i.e. for the gravitational forces to take over and the Universe to start contracting. Such a system would then shrink back to the original primeval cosmic egg and possibly undergo a Big Bang again.

The intergalactic gas revealed by X-ray astronomy seems to tip the balance in favour of a closed Universe and a future of shrinking and then recycling or oscillating until the energy, radiation and mass is completely lost in a weird form of cosmic heat death (see page 216).

I very much doubt if Roentgen — even in his wildest speculations — ever dreamed that his discovery might provide evidence that would be used to settle the fate of the Universe!

X-ray Optics Telescopes and Detectors
By now, several questions are probably being asked: if X-rays recorded in space are as penetrating as those produced on Earth, how do we focus them to produce a directional beam?; or how do we build an X-ray telescope and point it at various parts of the sky and thereby correlate an X-ray source with a visible object?

'How do we detect X-rays?' must also figure as a leading question, and it is the one most easily answered. The first X-ray detectors were simply very heavily shielded photographic plates, and this also partially answers both the previous questions, because the early X-ray telescopes were little more than modified pinhole cameras. Photographers are familiar with this technique which relies on the fact that a minute hole pierced in an otherwise opaque plate allows collimated light to pass through it. This forms an upside down focussed image on a back plane opposite the pinhole. Pointing the pinhole at a scene and placing a photographic plate on the back plane produces a fully representative image of the scene on the plate when the latter is developed and fixed in the normal manner.

Early X-ray cameras were of the pinhole type and were essentially nothing more than shielded boxes (which were opaque to X-rays) fitted with a precision-drilled pinhole and a shielding exposure shutter. Fortunately the early success in finding X-ray sources allowed the experiments further funding to proceed. These early sources were so powerful that they produced highly usable results from these very crude cameras. As the nature and energy levels of celestial X-rays were more fully understood, so more sophisticated focussing and detection systems were evolved. Photographs were entirely suited to work which lasted for only a few minutes (a rocket flight) and could be adapted to work on balloon-borne platforms. But for satellite work a more sophisticated system was needed, and not surprisingly the techniques used were borrowed from medical X-ray equipment and developed to suit the weight, volume, power, and environmental requirements of space technology.

The detectors now used are of fluorescent salts. These materials are usually combinations of the 'alkaline earth metals' — e.g., sodium, potassium, caesium, barium, strontium — and 'halogens' — i.e., fluorine, chlorine, bromine, iodine. These crystals flash or produce light when struck by X-ray photons. In several cases different materials are used for different energy levels. The X-ray astronomer is interested in 'soft' X-rays of about 0.1keV up to 'hard' X-rays of 100MeV or more. These are approximate figures: at the 'soft' end of the spectrum the energy levels overlap into the ultraviolet, while at the 'hard' end they overlap into gamma-rays. The soft rays have very little penetrating power and can be stopped by thin layers of metal such as aluminium. The hard rays are capable of considerable penetration and are stopped only by heavy shielding with lead. As a comparison, the X-rays used in medical work operate at about 50-150keV, which is roughly midway between the levels that interest the astronomer.

This difference in penetrating power has a considerable effect on the 'optics'; i.e., the focussing and collimating systems used to obtain narrow directional beams of rays which allow accurate position measurement, and the accuracies obtained are inferior to those for optical telescopes. At the low-energy end of the spectrum (0.1keV) a mirror may be used. It must be very highly polished and of a very high optical quality — the wavelengths are so short that the smallest scratch on the mirror surface would scatter the rays. Also, as the wavelength decreases, the mirror fails to reflect as the X-rays try to penetrate the material. Soft X-ray mirrors have been made from gold, aluminium, platinum, tungsten, osmium, indium, rhodium and rhenium with varying degrees of success. The reflectivity has been improved by the use of proprietary specialized coatings, but nevertheless seldom attains an efficiency in excess of 20%. This may be compared with the 80-90% reflectivity for optical and infrared telescopes.

The harder X-rays cannot be brought to a focus by reflection. Instead, the beam is reduced to a small diameter by means of a small hole in a shielding plate — i.e., a modified form of box pinhole camera. The simple pinhole is, however, carried a stage further. Instead of being just a hole drilled in a plate, a narrow-bore thick-walled tube is placed over the hole. This modification ensures that only those X-rays parallel to the bore of the tube and in line with the hole will penetrate into the box. This latter item holds not a photographic plate but a crystal scintillator. The number of light flashes produced is in direct proportion to the number of X-ray photons that strike the caesium iodide scintillator crystal. An electronic counter measures the number of flashes in a given time (say at intervals of 0.01 second, 1 second, 100 seconds) and records this number as an intensity level. If the X-ray source is a flickering nova or a cyclic X-ray pulsar then the count intensity will vary with time and in a sympathetic manner. By transmitting these intensity counts as a modulated telemetry radio signal to a waiting receiver on the ground, we obtain a record of the X-ray intensity of the object on which the collimated counter is trained.

This method is simple and accurate and, by sweeping the instrument across the sky and recording the counts, a map of the major X-ray objects will gradually build up. The work is painstaking and routine, but periodically an interesting object such as a pulsar or an X-ray flare star is found. When this happens efforts are made to try to identify the X-ray source with an optical counterpart. Here there are minor problems for the collimated tube system can be accurate only to a few tens of seconds of arc in terms of viewing angle. Theoretically the accuracy could be higher but this would demand a smaller bore down the tube, allowing fewer X-rays through and upsetting the sensitivity of the counting system.

The disadvantage of inaccurate X-ray measurement can be reduced by using more than one satellite's measuring system. Where the error circles or boxes overlap is the area where the optical counterpart is most likely to be found. Sometimes there may be ten or more stars in the error box — and then every one must be photographed several times and carefully checked for telltale signs. Usually the astronomer is looking for very hot blue or violet stars which flicker in an irregularly variable pattern or which are pulsating regularly in sympathy with the X-ray count. Cygnus X-1 was found in this way, as were several other sources.

More recently the X-ray pulsar in Norma was found by a telescope fitted with an image intensifier made by EMI Ltd. This instrument 'magnifies' or 'multiplies' the faint dribble of light photons emitted by the pulsar and allows the taking of a photographic or videotape record. This device may still be regarded as in its infancy, for a development of it may allow us to 'see' even fainter objects than can be directly recorded at present. By using a 'window' of aluminium of a known thickness an imaging tube could be penetrated only by X-rays of a certain minimum energy level. By allowing these rays to fall onto a screen coated with caesium iodide, a visible light image would be formed. This could be intensified in the normal way and recorded.

Uhuru was the first X-ray satellite and allowed four X-ray source survey catalogues to be completed before its transmitters fell silent. These surveys were catalogued and categorized as 1U, 2U, 3U, 4U, so that 4U 1956 + 35 is an X-ray object found on the fourth Uhuru survey at Right Ascension 19.56 hrs, declination + 35°; in fact, it is the catalogue number of Cygnus X-1.

Uhuru was followed by Copernicus (OAO-3) which, launched on 21 August 1972, carries three X-ray detectors designed to deal with various expected energy bands and is still functional, as is Ariel 5, which has probably done the major work in detailing X-ray sources. Now the task is being taken up by the HEOS (high energy orbital satellite) series, the programme of which is scheduled to carry on through the next decade.

6 Chips in Space

In 1952, four years after the invention of the transistor had been announced, Geoffrey W. Dummer, a British scientist then working for the Ministry of Defence, prophesied that the time would come when whole working circuits would be made from blocks of semiconductor material. Dummer had been working on the problems of miniaturizing electronic equipment and considered that the real revolution in electronics was yet to come.

He has been one of the few prophets who have lived to see their forecasts completely vindicated. What Dummer called 'blocks' of semiconductor material are now known as 'chips', and originally any system made in this manner was called a 'solid-state' circuit — nowadays it is called an 'integrated' circuit or IC.

The arrival of the silicon chip has ushered in a new age, and it is extremely difficult to assess how much of an impact will be made by this technology. Small data-processing chips are invading areas which were the stronghold of motor-driven timers. These are replaced by counters which divide down and count the incoming mains frequency and produce output pulses in seconds, minutes, hours and days. A digital watch or clock is now a commonplace piece of equipment. It is silent and has no moving parts except the electrons rushing around the components in the circuits. Integrated circuits have mated with optical items to produce optoelectronic displays using light-emitting diodes and liquid crystal materials. In television, the micro hand-held TV camera, using a single selfscanning photosensitive chip, is already a reality and is commercially marketed.

Dummer thought that a solid circuit might be a reality in 'about 20 years'; i.e., around 1972. In fact the first true solid-state circuits were produced and marketed, by Texas Instruments, in 1959, and this breakneck pace has continued without showing any signs of slowing down. This first integrated circuit was a simple divide-by-two stage: it had two transistors, four diodes, four capacitors, and six resistors; i.e., a total of sixteen

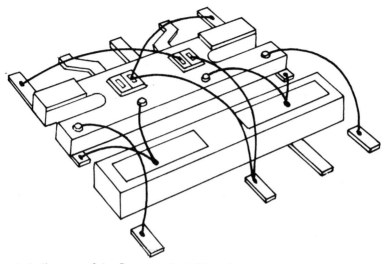

6 A diagram of the first practical silicon integrated circuit.

7 The circuit diagram for it. With a total of 16 components the circuit was a simple binary divider or 'flip-flop'.

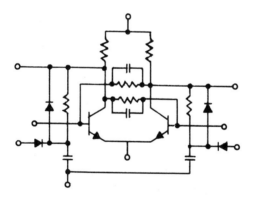

components fabricated on two blocks of silicon. The complete assembly was mounted in a ceramic-metal package about 0.75cm square and equipped with ten leads to the outside world. Since the whole assembly looked like an insect these packages were soon christened 'bugs' — a name that has stuck through twenty years of intensive development.

They have grown with frightening rapidity. In 1959 a package had ten legs and housed sixteen components. By 1974 they had forty legs and housed *11,000* transistors on a flat chip of silicon 0.5cm square; now (1979) they house *150,000* components on the same size of chip and have up to 64 legs on the bug package. By 1985 the manufacturers confidently expect to place *1 million* transistors on a single chip and interconnect them to form an extremely powerful microprocessor — literally a large computer on a chip. Such complete chip computers are already available: the illustration in Plate 2 shows such an item, a smaller-scale (in terms of computing power) version of the 1985 article.

I once asked the technical director of a major manufacturer of integrated circuits just how far we were from the final limit of miniaturization. His answer was sobering. 'We have a long way to go,' he replied. 'Just nine atoms' thickness of silicon is all we theoretically need for a transistor; a three-atom-thick layer of "n-type", a three-atom layer of "p-type", and a final three-atom layer of "n-type" will give us an adequate n-p-n transistor. Since we are running at thicknesses of microns [10^{-6} metres] we have between 10^4 and 10^6 times to go before we hit rock bottom!'

I stress this point because I so often hear (and see) people say 'we have a long way to go' while tapping the head, meaning the human brain has a far higher component-packing density than present integrated circuits. This is true at present but, since the brain is made of long complex organic molecular chains, its theoretical packing limit is far lower than that achievable with silicon. As development proceeds, I think we have every chance of fabricating silicon 'chips' which will rival the brain in complexity — and possibly eventually exceed it. The silicon circuit's history, from Dummer's prophecy to the million-transistor chip, spans just over thirty years, while the human brain has been developing for a few million years. Silicon *has* got a 'long way to go' but it certainly is travelling in the right direction.

Computers rivalling the brain in capacity, although not in intelligence, will probably emerge by about 2000AD — if not before. A true silicon artificial brain of the right size and intelligence may not appear for another century — I would, therefore, expect such a device to materialize around 2300AD but

I could be just as wrong and pessimistic as was Dummer!

It is not generally realized that spaceflight requirements initially *forced* the development of integrated circuits. The need to pack extremely complex gear into small spaces and guarantee a reliable working life for the whole assembly demanded such techniques. Now of course the IC industry is well able to look after itself.

Silicon circuits have made the greatest single contribution to the success of astronomy above the atmosphere. Without them, the reliable X-ray observations from Ariel 5 and OAO-3 Copernicus would not have been possible. The reasons are varied but the following are major:

a) The launch rocket would have been enormous and beyond the development capacity of existing technology. There was just no alternative to IC development.

b) The power demands would have been prohibitive — valve (tube) equipment consumes hundreds of watts. Modern ICs consume milliwatts or (in some cases) microwatts.

c) The equipment would have been unreliable and would have either failed prematurely or demanded some form of planned maintenance. Arthur C. Clarke had this point in mind in his original designs for space stations.

Today ICs can be guaranteed to give working lives of thousands of years (based on accumulation of enormous quantities of data on test results and actual working hours). The fantastic reliability of an IC is due to the enforced cleanliness required during manufacture, and the use of metallic-silicon interfaces that are truly compatible. Joints or connections on a silicon circuit are either buried deep in the silicon itself or layered on the top of the chip in the form of an aluminium network. Multiple layers of interconnection may be used, and the conductor widths are presently 10 micrometres, although 0.5 micrometres are promised for the million-transistor chip.

Although silicon circuits are complex, they are not costly and can be (and are) mass-produced to very tight tolerances. The tools used to manufacture them are photographic and the dimensions and tolerances are entirely controlled by the wavelength of light used to make the photographic masks. Originally exposed by visible light, they are now prepared by ultraviolet light; since this is a much shorter wavelength, the diffraction effects are reduced, the fringes surrounding the images are much less, and consequently a sharper and more clearly defined picture is obtained. This allows more circuitry to be photographed onto a given area of silicon, and smaller interconnection patterns to be established. Logically, from our knowledge of the electromagnetic spectrum, we would expect 'soft' X-rays to produce even sharper images, and this area of the spectrum is now being exploited to produce even smaller patterns.

As an example, if we expect to diffuse 10^6 transistors into an area where we could previously only lay down 10^3 then the transistor density per unit area has obviously gone up by 10^3. Since the chip is probably square this means a linear increase in density of about 31. If we originally used yellow-green light of 0.5×10^{-6} metres (500 nanometres) wavelength for definition of 1,000 transistors/square then we need

$$\frac{0.5 \times 10^{-6}}{31}$$

or 1.6×10^{-7} metres (16 nanometres) wavelength for 1 million transistors/square. This is definitely a 'soft' X-ray region of the spectrum. The tools or masks for photography at these wavelengths are thin metallic foil. As experience is gained in such processing it is reasonable to expect that shorter X-ray wavelengths will be used. The limit is finally set by the thickness and material of the photographic mask. If the X-ray wavelength is too short the mask is penetrated willy-nilly and no longer effectively defines the masked area; the mask effectiveness would then be controlled by choice of material, lead being a better agent than (say) aluminium. The limit is also set by the difficulties of satisfactorily focussing the mask pattern onto the surface of the silicon base material.

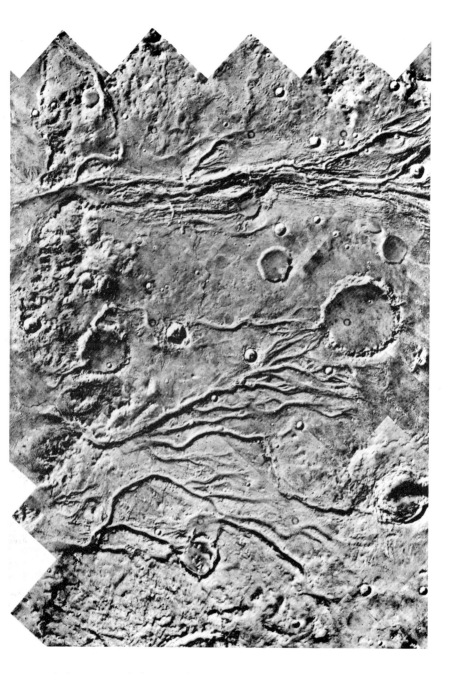

11 A close-up mosaic from the Viking 1 orbiter of the Martian surface. Clear in the illustration are channels thought to have been formed by flowing water – something thought never to have existed on Mars before evidence such as this from the Viking mission. (*Courtesy NASA.*) See page 91.

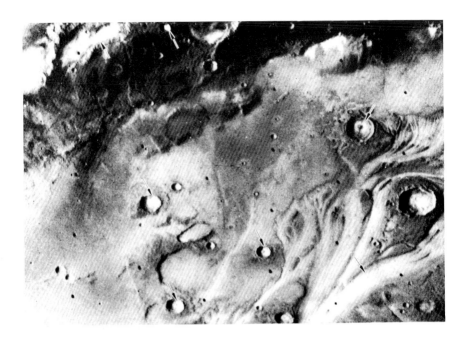

12 & 13 Two pictures, taken a half-hour apart by the Viking 1 orbiter, show the development on Mars of early-morning fog in low spots. (*Courtesy NASA.*) See page 91.

Manufacture of a Silicon Chip

The basic process, planar diffusion, by which a chip of silicon is turned into a circuit performing logic functions is a very simple one. All the agents and materials are diffused or percolated through one side of a thin slice of silicon. The process was made possible by the discovery that silicon oxide is impervious to the 'dopants', as these agents are called. Therefore the slice of silicon is first coated with oxide by the simple process of blowing steam over red-hot slices of silicon of about 10 to 15cm diameter and 0.1mm thickness. The growth rate of the oxide skin is carefully controlled by the temperature of the slices and duration of steam exposure.

The oxidized cool slices are then coated on the top side with a photo resist material. This is photosensitive and, wherever it is exposed to and illuminated by light, uv, or X-rays, it will harden. The unexposed material is soft and easily washed off with water. This leaves exposed areas of the oxide which correspond with the contours of the mask pattern. The slices are now dipped into an etching solution (usually based on hydrofluoric acid) which dissolves the exposed areas of oxide down to the basic silicon. The hardened photo resist is now removed by a suitable solvent and we have the desired transistor layout pattern etched in oxide on the surface of the silicon.

The etched slices are then placed in a furnace and heated to about 1,300 K, and a gentle stream of gas containing the required doping agent is blown over the assembly. Wherever the silicon is encountered, dopant enters and changes the silicon to p- or n-type material, depending on whatever dopant is used. In this way, by successive operations of masking and etching, the base and emitter patterns of transistors are superimposed and the p-n junctions of diodes are built up, and if required the patterns of resistors and capacitors deposited.

Finally, an interconnection pattern of aluminium (also determined by photo resist etch) is vacuum-deposited on the finished circuits to connect the components to form the desired logic or computer function. The slice is then cleaned, washed, dried and tested. Only the good circuits are required, so any defects are 'squirted' with a distinctive coloured dye for rejection at a later stage. The slice is then *very* carefully scribed with a diamond and broken into squares.

Each square chip contains a complete circuit and, since each slice may carry about 150 circuits and a batch of slices may number about 20 to 50, between 3,000 and 7,500 circuits are made in one batch. The 'numbers game' now mounts up swiftly. From raw slice to finished circuit takes about 8-12 hours and, on a shift basis, two batches or about 15,000 circuits would theoretically emerge from a simple 'work through' scheme. But the manufacturer works on a continuous production basis, and it is not unusual to have several furnaces 'teamed up' to meet requirements.

The diced chips are now individually sorted, the dyed chips being rejected and the good ones retained. Manufacturers are coy about reject levels but at this stage 60% to 80% of the circuits may be rejected — they cannot be reworked and are scrapped completely. Daily production from a typical area may therefore be about 3,000 good circuits.

But if each circuit contains 1 million transistors, then the *daily* output of the system is the equivalent of *3 billion* transistors — something that even the very best automated transistor production area would find very difficult to match. This output would be even more difficult for the valve manufacturer to achieve. It explains why 'computers on a chip' are feasible and are being made economically and reliably as a consistent product.

For satellite use, even more stringent requirements must be met and, in addition to extra-rigorous testing, the manufacturer enters on special methods of manufacture (much the same as those described above; details are beyond the scope of this book). In addition to surviving the hazards of launching, integrated circuits must also survive nuclear radiation doses from various items such as the Van Allen belts of Earth, similar belts around other planets (Jupiter is a good example), and solar storms during periods of intense sunspot activity.

Silicon chips therefore have given us the much needed reliability, small volume, low power consumption, good cost-effective product desired in space vehicle equipment. There is simply no alternative technology which could have made this possible, for the use of a microprocessor on board a spacecraft confers a high degree of flexibility and a certain amount of intelligent decision-making in the system. It is able to think for itself, and as silicon circuits become more sophisticated so the pressure is developing to produce matching software.

Flexibility is Software!

In short, there are still several avenues of development left and these include improving software; i.e., computer languages. The operational flexibility of a microprocessor-controlled system lies in the adaptive powers given by the programme memory and software instructions; merely by changing key instructions a whole new set of operations may be carried out *by the same equipment.*

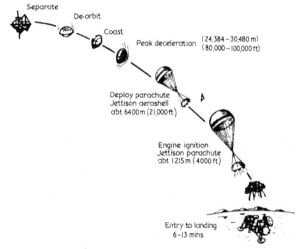

8 The entry sequence for the landing on Mars of the two Viking space-craft. Some of the photographic results of this mission can be seen in plates on pages 70, 105, 106 and 123.

A good example of this adaptive quality is seen in the Mars Viking missions. These craft carried microprocessors on both the landing craft and on the orbiting craft circling the planet. In the launch and flight phase of these missions the two computers were linked and 'talked' to each other over solid wires. When the landing craft separated, the two computers went into different operating modes. The lander craft computer controlled *all* the phases of atmospheric entry and touchdown. When settled on the surface it commanded the extension of a radio link antenna and the commissioning of several separate experiments. The orbiting craft 'listened' for the link signals and, when located satisfactorily, sent the good news back to Earth: 'Vikings are OK.' The computers, previously linked by wire, were now linked by radio and carrying out two separate and completely different routines.

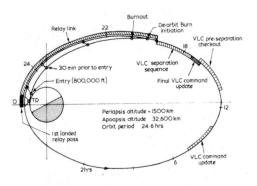

9 Typical separation and landing orbit for the Viking spacecraft on reaching Mars (VLC refers to the Viking lander).

Not only that, but when the 'scoop' of one of the vehicles was jammed a new set of instructions was written and sent to Viking which, following the new routine, finally freed itself and functioned as intended. And the Viking may safely be considered a dunce or idiot in comparison with the later mission vehicles being planned and constructed. These language routines can be expected to become even more complex with instructions approaching the million-plus mark for really deep-space (beyond Pluto) vehicles. A semi-intelligent Mars Rover is shown in Plate 19.

Silicon circuits have now reached the stage where they can be made self-repairing! If a short or open circuit is detected then the system is bypassed or a new one switched in, or the old one repaired by allowing the heat developed to fuse metal which solders the break.

Man has not yet ventured very far into space but he certainly is sending out unmanned 'vehicles in his name', judging by the way they work. Many people may consider these developments as 'inhuman'. Nothing could be further from the truth. In the way they work, and in the manner of their interaction, they bear the hallmark of no one but Man, even although it be written in metalwork, plastic, and silicon chips!

7 The New Optics

If the 'average knowledgeable person' were asked 'Where is the world's largest optical telescope?', he or she might answer: 'Mount Palomar, in the USA' — and would be wrong! A small and more knowledgeable group might answer: 'At Zelenchukskaya in the USSR — and would be technically right. But a smaller group might reply: 'At Mount Hopkins, the new multi-mirror telescope (MMT).' This group would be both right and wrong. For the MMT is *physically* larger than the 6m Zelenchukskaya installation, but its actual *collecting* area is smaller. The engineering principles used in its design are unique, for it is the optical astronomer's first attempt at obtaining very large 'equivalent apertures' by means of phasing or combining the separate outputs of several smaller mirrors. This instrument is a pilot scheme for a much larger 'mirror' which will have an aperture equivalent of 25m. This would easily make it the world's largest telescope for it would effectively have 25 times the collecting area of the Palomar instrument.

This clearly indicates how the new disciplines of radioastronomy and spaceflight have been turned round to improve the characteristics of ground-based instruments. The technique of optically phasing or combining even two separate images would not have been considered feasible twenty years ago, but the Arizona University instrument at Mount Hopkins combines the images from six separate mirrors each 1.5m in diameter. Obviously great strides have been made in the centuries-old art of grinding and polishing mirrors to extremely fine limits. Also, there have been developments in the maths of designing mirrors — or, more important, *mirror combinations* such that the overall performance of the system is not so seriously disturbed by minute position changes in these mirrors.

These developments in ground-based telescopes can be directly attributed to work done specifically for space items. A space telescope is subjected to extremes of temperature: the Sun may be shining on one side while the other side faces the 3 K background of space, and so the telescope's supporting framework must change shape or flex as little as possible — in a known and controllable manner. If this is done then the 'ray path' — i.e., the trajectory the light beams follow — can be altered to compensate.

Space telescope mirrors are made of fused synthetic quartz or silica or of special glass ceramics called 'Cer-Vit' or 'Zerodur'. All of these materials have an extremely low temperature coefficient of expansion; i.e., they will not crack or change their focal length or range due to temperature changes.

An even more remarkable material is being used for the NASA Space Telescope due to be carried into orbit by the Shuttle in about 1983-84. This 2.4m mirror is made from a specially formulated Corning mixture officially known as Code 7971 but popularly known as 'ULE' (ultra low expansion), the major ingredient of which is titanium silicate. It is a truly zero-temperature-coefficient material.

The normal way of making a large telescope mirror is to cast it from the required glass after the latter has been heated to a molten state. Casting is a lengthy and costly process (the Palomar mirror was cast three times before a flawless result was obtained), and even then it results in a heavy object which must be handled carefully during the lengthy grinding and polishing process.

The mirror made by Corning from the ULE 7971 titanium silicate was fused together from selected plates and struts using a carefully controlled hydrogen welding torch. The process may be likened to modern methods of making welded iron or steel structures when compared to the nineteenth-century method of casting them: the modern structure is extremely light and yet is far stronger than the cast version. So it is with the mirror welded together from plates and struts of titanium silicate. Materials like synthetic silica, 'Cer-Vit', and ULE 7971 are less time-consuming in grinding and polishing. By the use or semi- or fully-automatic methods of measurement, the process of grinding, polishing and figuring is being cut down to a year. The 5m 'Pyrex' glass mirror for Palomar (again produced by Corning) took over four years to cool, grind and polish to its final perfection!

The 1.5m mirrors for the MMT are also produced by largely automatic processes and, since they are of a 'standard size', can be produced at once on six separate machines. Thus the overall combination of automated production of compensated ray path mirrors with zero temperature coefficient, housed in a compensated support frame, has produced a unique, low-cost instrument of superb performance.

Two normal 4m mirrors have recently been made and installed, one at Kitt Peak, Arizona, the other at Cerro Tololo in Chile. The UK has recently installed the 3.9m mirror at Coonbarabara, Siding Springs, New South Wales, and this superb instrument is *also* made of very low-expansion 'synthetic' glass. Other similar telescopes are being installed in other key areas of the world. All of them can trace their technological perfection to the need for good optical telescopes in space.

The projected 25m aperture is a logical extension of these developed techniques. Four design concepts are being considered: a steerable dish surfaced with mirrors; a segment of a sphere or 'shoe' surfaced with mirrors; twelve individually mounted telescopes arranged so that their beams can be phased and harmonized; and, fourth, a set of small mirrors on a steerable frame. This last concept is used by the Johns Hopkins team. If we assume that the six-mirror combining system is repeated on a larger scale, then each mirror would need to be 10.2m in diameter for an equivalent total area. This is twice the size of the Palomar mirror, and is probably beyond the manufacturing capability of the foreseeable future.

If we consider the very best large 'mass-produced' mirrors — the 4m Kitt Peak and Cerro Tololo items — then in order to produce an equivalent area we must be able to align and phase no less than 39 such items! This will place heavy constraints on the accuracy of construction of the supporting framework. Because they are relatively easily mass-produced, the aim is to use smaller mirrors of 1 to 2m diameter. Using 2m items would require approximately 160 mirrors, and using 1m diameter we would need 625 reflectors. Although the phasing problems are thereby increased, they are not insoluble.

If one mirror is wrong or out of phase it can be shut down. If this happened, only 0.16% of the light collecting ability would be lost — in fact, if this giant were to be commissioned it is doubtful if every one of the 625 mirrors would be functional at any given time. Indeed, such a system could be kept *continuously* in service by adopting (say) 650 mirrors and stripping down and realuminizing them a set at a time (batches of twelve or thirteen on a weekly basis would make up the year). Observation with large single-mirror telescopes would be temporarily halted when the glass is re-aluminized.

Furthermore, if a small mirror developed a fault it could be replaced by a good one at almost any time. Not so with a single mirror: if any flaws develop they have to be permanently blanked off; the huge Soviet 6m mirror has 20% of its surface area blanked off because of such defects. Two more 6m mirrors made of the Soviet equivalent of Cer-Vit are now being manufactured for installation as soon as possible. A 20% loss is serious in terms both of light gathering and of reduced resolving power.

'But why build a giant 25m telescope *on Earth?*' is a fair question to ask, and the answer is provided by Dr. Leo Goldberg, Director of Kitt Peak. 'The aperture is large enough to permit detailed spectrographic research on winds and weather on the other planets in our Solar System.' It would also (adds Dr. Goldberg) allow accurate measurements of the diameter of Pluto. Many of the smaller moons and satellites, and possibly comets, too, would be accessible to measurement. We also might be able to directly detect several new faint stellar bodies (nearby white dwarfs) and even the suspected planetary companions of nearby stars.

To this I would add the ability to measure the actual disks of other Sun-like stars by means of speckle interferometry and (using the same technique) the larger stars. We should, for example, be able to measure starspots on Sirius or Alpha Centauri. We may even find that some classes of star do not have such features, and conversely we may expect other stars to have exaggerated spot and flare characteristics. A precision scan of Proxima Centauri (the nearest star known outside the Solar System), which is a well documented flare star, could be extremely informative. Speckle interferometry photographs may well show large flares but probably will not show the actual spot or flare site.

a	b	c
Speckle photograph Single dots	Speckle photograph Single lines	Speckle photograph Paired dots
d	e	f
Fourier spectrum of a point	Fourier spectrum Partly resolved	Fourier spectrum Fully resolved
g	h	i
Final image Unresolved point	Final image Partly resolved binary	Final image Fully resolved binary
UNRESOLVED OR SINGLE STAR	PARTLY RESOLVED BINARY INDICATING POSITION ANGLE	FULLY RESOLVED BINARY INDICATING POSITION ANGLE

10 The assembly stages in forming Fourier speckle interferometry images. For full explanation, see text.

Speckle Interferometry

Elsewhere the resolving power of a telescope has been expressed by the formula $r = 1.2\ \lambda\ /D$, where r is resolution in radians, λ is wavelength, and D is the diameter of the reflector. If, then, we are examining a star with a telescope with an aperture of 25m (if we assume a centre band wavelength of $0.5 \times 10^{+6}$m) the resolution becomes $1.22 \times 0.5 \times 10^{-6}/25$ or approximately 2.5×10^{-8} radians. One light year is approximately 10^{16} metres, so the minimum resolvable size of star at 4.2 ly (about 1.3pc) is given by

$4.2 \times 10^{16} \times 2.5 \times 10^{-8}$m; i.e., approximately 10^9m or 1,000,000km. This is about 0.75 times the diameter of the Sun and about five times the diameter of Proxima, so it is very doubtful if such a tiny star would be resolved properly by a 25m aperture telescope, but the giant flares it produces may be resolvable.

The interference produced by the atmosphere will not allow a resolution approaching even 1% of this value; resolution by ground-based telescopes is limited to objects separated by about 1 arc sec — as opposed to 0.005 sec, the true theoretical limit. However, the technique of speckle interferometry which we have mentioned allows an image quality and resolution approaching theoretical limits even with the atmosphere's 'interference'; in fact, it is this very property of the atmosphere that is exploited in speckle interferometry.

The atmospheric shimmer is caused by the lower layers forming bubbles or cells of differing density and refractive index. These bubbles, about 10cm in diameter, each produce an image of the star being examined. But, as the bubbles disappear and reform continuously, the multitude of star images produced melt together and form the fuzzy outline that characterizes an attempted high-definition photograph taken through the atmosphere. If we could take a very fast photograph which 'froze' the motion of these 10cm bubbles then each of the cells would contain a fuzzy image of the star being photographed. But, although poor in quality, all the data and information necessary to produce a high-quality image is contained within the photograph.

The images are said to have been 'convolved'; i.e., rolled up or mixed or distorted by the defocussing properties of the atmosphere, the graininess of the film emulsion, and the optical distortion curvatures of the telescope and camera. If we can 'deconvolve' these distortions we should theoretically get almost perfect images. Deconvolution is a trick now performed by computers using specialized algorithms (trial-and-error routines which give a high degree of accuracy). These are able to deconvolve or unscramble the twisted up information and piece it together correctly.

One *very* powerful technique for doing this photographically is to take a series of very short exposures and develop the negatives carefully. Then, from *each alternative negative,* a positive transparency is produced and carefully dimensioned to be the same size as its predecessor. If we had ten photographs taken in rapid succession and left 1,3,5,7,9 as *negative* but printed 2,4,6,8,10 as *positive,* then by superimposing 1 on 2, 3 on 4, etc., the wanted permanent sections are cancelled out but the *distortions* are left in the form of odd shaped black or white sections of the images. In effect, what we have done is to separate the signal information (the star image) from the noise information (the unwanted stuff).

The next step is to subtract the noise from the original photograph, and this can be done by superimposing a negative noise picture on a positive original. The noise is cancelled out and the noise-free image is left. A further improvement is effected if we also superimpose a positive noise picture on a negative original. We now have two pictures (one positive, one negative) which are almost noise-free but possibly with a few traces of imperfection due to minor changes which occurred between photographs. If we then invert *one* of the pictures and superimpose it on the other, the permanent sections of the images will be reinforced. What we have done in effect is to separate the noise from the wanted image, then inverted the noise detail and subtracted it from the original to obtain a clear image. Superimposition of these pocessed images produces an even clearer picture. This process may be continued until no further improvement is noticed.

Nowadays the process of inversion and subtraction is done electronically by computers using algorithms involving Fourier transforms. By using Fast Fourier transforms (FFTs) the process may be done in real time at television picture rates (50-60 frames/second).

I have deliberately described the photographic system, for it explains the process in everyday terms — and actually *is* used to obtain certain types of deep-space photograph. If the star is single, or a binary beyond the telescope resolution limits, the series of 'speckles' obtained resembles a cluster of dots; if the star is a barely resolvable binary these dots are elongated; but if the image is resolvable by the telescope then the dots appear as separate pairs. As stated, each dot or speckle represents a miniature image of the star as seen through an atmospheric cell or bubble, and the experimenters at this stage are immediately able to see if they are dealing with a binary system, merely by looking for elongation or separation of the 'speckles'.

If the speckles are illuminated with laser light, in the resultant light pattern (the power spectrum) the individual speckle separations now line up to form interference fringes. In a further stage, termed speckle holography, the power spectra may then be further processed to yield full images, either single or resolved. The speckle system may be used for resolving not only the separation of binary stars but also stellar surface features. The first star to be so revealed was Betelguese (α Orionis), the brightest star of Orion. The fully processed feature-enhanced image shows what appears to be a seething mass of convection currents with some objects looking like giant sunspots. 'Giant' is not used in a superlative context for the smallest 'spots' appear to be about 10^7km in diameter: this is about six times the total diameter of our Sun, which could be dropped into one of these disturbances and vanish without trace.

Several star systems have now been resolved. A French team have examined one hundred stars and starlike objects including Algol (a well known eclipsing binary), Arcturus (α Boötis and Nova Cygni 1975. In the latter case the observers have been able to resolve the shell of glowing hydrogen surrounding the star 45 days after the outburst. The observers conclude that the method works on present telescopes for stars down to the 13th magnitude. It could obviously be applied to the 25m aperture, which (with improved electronics) would allow resolution of objects down to 18th or possibly 20th magnitude.

The performance of such systems will only be bettered by the Large Space Telescope and possible Lunar Observatories. But even these may need speckle interferometry equipment to handle the scintillation from interplanetary and interstellar dust clouds.

I would make the point that speckle interferometry is *extremely* useful if the axis of revolution of the binary is in the plane of the telescope. Under these conditions a close binary cannot be spectroscopically resolved, for there is very little spectral Doppler shift obtained. On the other hand, if the axis is at right angles to the plane of the telescope, the speckle system will *still* resolve the images if the separation is within the Rayleigh resolution limit. Binary systems that are otherwise known spectroscopically, or previously thought to be single stars, thus become resolvable.

An astonishing demonstration of the power of speckle interferometry has been the discovery that Epsilon Eridani is a binary. This object is one of the nearest sun-like stars, of type K_0 with 0.3 of solar radiation, 0.8 solar mass, and lying at a distance of just over 3.7 parsecs from the Solar System. This star was one of the two candidates examined by Frank Drake in the 1960 Project Ozma, the first experiment aimed at listening for intelligent radio signals from a possible alien civilization.

The presence of a large planetary companion (about six times the mass of Jupiter) had been suspected by Van de Kamp (see Chapter 8), but in 1976 Blazit and his team of the Observatoire de Paris, Meudon, using speckle interferometry, resolved the star into two components separated by about 50 milliseconds of arc. The position angle is quoted as $143 \pm 6°$ (but in speckle interferometry resolution there is an ambiguity of 180° and so the position angle could equally well be 323°). The orbits of the binary components cannot yet be properly determined and further observation is required.

So, as of 1976, Epsilon Eridani is known to be a close binary and, accordingly, unlikely to be a life-supporting area — the radiation extremes would seem to be too great. It is not yet known whether the perturbations recorded by Van de Kamp are real and due to a massive planet. If so, it could be an unusual 'solar system' configuration of two K_0 type stars only 0.16 AU apart orbiting each other in a period of about 20-30 days, the pair being orbited in turn by a planet six times the mass of Jupiter with a period of 26 years and a separation distance from the double primary of about

8 to 10AU. But who are we to judge what is usual or unusual — we must expect every possible combination of star/planet configuration to occur *somewhere* in the cosmos!

Multiplying Light

We have seen how the astronomer has hopes of building optical telescopes with large effective apertures by assembling several smaller mirrors and arranging the collected light images such that they are superimposed in phase. Eventually, however, should the astronomer wish to see fainter objects, this technique becomes too costly and the instrument too large.

But our astronomer can turn to electronics and ask the engineers in this discipline if they have any answers. They have. The solution to the problem is a technique known as image intensification. It is a development of an earlier technique known as photon multiplication; the instrument performing this task is termed a photomultiplier. If a photon of visible (say blue) light is caused to fall onto a metallic silver plate or cathode coated with silver oxide and caesium, then as the photon strikes the surface an electron will be emitted. If a plate is placed adjacent to the cathode and biased with a positive voltage, the emitted electrons will be collected by it; the resultant current will be an analogue of the light signal falling on the cathode. This is a straightforward photoelectric cell. If, however, this anode is itself coated with caesium/silver oxide, the arriving electrons will cause the emission of further electrons. And if a further coated anode is added, it too will emit yet more electrons ... Each stage will act as a multiplier and add its quota of electrons to the final output.

The overall gain or multiplication achievable in a complete multiplier comprising (say) eleven stages is about 10^7. One million photons per second arriving at the cathode are converted to 10^{14} electrons per second (this is entirely dependent on electrode voltage supplied); since an electron flow of 10^{19} per second corresponds to a current of 1 amp, our photomultiplier tube will produce a current of 10 microamps at the anode. Currents of this level are easily amplified for use as (say) television signals.

But the photomultiplier could be mounted at the focus of a telescope mirror, and thus amplify any very faint light signals detected from pulsars or other objects. In theory, therefore, we could look at objects with our telescope 10^7 times fainter than would be possible without the photomultiplier: this corresponds to a magnitude decrease of about 17.5. If the photomultiplier were mounted at the focus of the 5m Palomar telescope, which has an unaided limit of about magnitude +23, the instrument would in theory be able to see stars down to about magnitude +40.5.

But things are not that simple! Thermal noise would have interfered with the system long before that *very* low level was obtained — magnitude +40 corresponds to seeing Earth-like planets at distances of three parsecs or so.

Photomultipliers are often used to obtain fast-event photographs *via* television systems — the Crab pulsar was photographed in this way — but the *real* imaging potential of electronics is realized in the image intensifier. In this apparatus the complete image available in a large telescope is focussed onto a photocathode of large area (typically 8-10cm^2). The *whole cathode* now emits an electron image which in spatial terms is an exact duplicate of the original light image. This is accelerated towards a coated anode, as in the multiplier system, but the electron image is kept spatially focussed. When this beam strikes the anode it gives a faithful reproduction in light of the original image, because the anode coating fluoresces, or glows, under the bombardment from the cathode electrons. These cathode electrons are more energetic than the original photons, and so their image is more intense. If this second image is focussed onto a further photocathode the image is even more intensified.

Intensifiers are made with up to four stages and can produce a photon-electron gain of up to 10^8. EMI specialize in the manufacture of these high-gain four-stage systems used solely for astronomical purposes, and as a result are able to boast participation in several 'firsts', the most notable recent achievement being the identification of the Vela pulsar (see page 16).

14 The first photograph ever taken on the surface of Mars, by Viking I minutes after landing. The slight blurring to the left is caused by settling dust in the wake of the craft's landing; because this photograph was produced by vertical scanning from left to right, the end-result of such settling can be seen in the form of the dust that has settled into the centre of the lander's foot on the right. (*Courtesy NASA.*) See page 91.

15 A spectacular photograph from the Viking I lander of the Martian horizon. (*Courtesy NASA.*)

16 A summer day on the planet Mars, as seen from the Viking 1 lander. (*Courtesy NASA.*) See page 91.

17 A visualization of the Viking lander. (*Courtesy NASA.*)

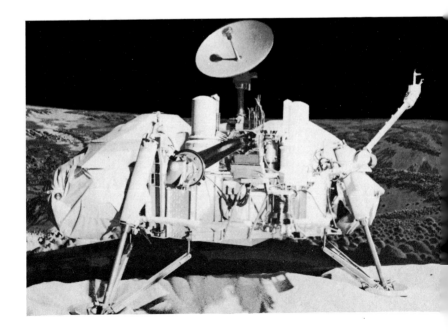

If used at full gain, such a tube needs to be cooled, as otherwise the stray electrons emitted by the warm metal of the tube would produce too much background noise. By cooling the tube to 253 K (-20°C) the noise is held to acceptable limits and the sensitivity of the system is then set by the light emitted by the atmosphere itself. See Plates 3 and 4.

Image intensifier tubes are being fitted to the telescopes carried aloft by the Shuttle, and eventually they will be in operation with the 2.4m Large Space Telescope. Without the limitations imposed by the atmosphere and with an overall gain (including electronics) of about 5×10^7 that of Palomar, the planets of nearby stars should be visible.

8 Radioastronomy

If the heavens of the optical astronomer are dusted with starlight then those of the radioastronomer are littered with noise. This 'noise' is created by naturally produced radio waves received *via* an aerial or antenna, amplified by a radio receiver, and then made to operate a loudspeaker or — more likely — a pen recorder. Most of the noise sounds like the hiss of escaping steam or the roar of an aircraft jet exhaust at full power. Some special noises sound like ancient breathless steam engines while others are more like rusty hearth crickets.

The radioastronomer has learned to decode and read these noises and deduce certain properties associated with the objects producing them. His position is closely akin to the doctor who must diagnose heart, chest or lung complaints through the sounds they produce. The radioastronomer is, indeed, in a slightly worse position, for his art is only about forty years old — compare this with centuries of optical experience.

The radio student now knows that there are several types of celestial radio transmitter that will never be seen in the true sense. They are too dim to be visible in even the most powerful telescopes — because they are too cool, too small, too far away, or too gaseous and thinly dispersed. But, just as the medical doctor is enlisting electronic diagnostic tools, so the radioastronomer — already by definition an expert in electronics — is now using sophisticated electronic systems to catch the faintest whisper of noise, isolate it from the background of manmade and unwanted other natural noises, and process it to extract a desirable feature and catalogue the results. Eventually the catalogue becomes a long list of noise sources, and after a while the noise varieties can be and are classified. In several cases the radio object so caught in the skynet is identified with a particular visible object, well known or otherwise. Classification invariably produces groups of objects with particular features — perhaps the radio 'pulsars' are the most familiar — but there are others which correspond to galaxies or peculiar stars in our Galaxy. Some of these objects are the remnants of supernovae which exploded in prehistoric times while others represent protostars about to collapse to form normal stars.

Recently, radioastronomers have taken to active sounding. Radio waves have now been bounced off the Sun and the Moon, all of the nearer planets, and out as far as the rings of Saturn. The results of this planetary radar work have solved puzzles that defeated optical astronomers. An astronomical unit (the distance from Sun to Earth) is known to within a kilometre, as are the orbital distances of all but the outer planets. The surfaces of the nearer planets have been scanned, and the secrets of the hidden face of Venus and that planet's rotational period as well as that of Mercury are now revealed. Over the last forty years radioastronomy has grown from a weak, squint-eyed infant into a sturdy multifacetted adult fully capable of aiding and abetting its older optical parent and — as often happens — sometimes exceeding the older generation's capabilities. The pace of growth shows no sign of slackening off, and there is every reason to believe that new discoveries will emerge from radio installations above the atmosphere.

The radioastronomer already has a window in the sky, inasmuch as the atmosphere is transparent to radio waves from about 20mm up to about 20m. Below 20mm the atmosphere molecules absorb radiation and, except for a few very narrow slots around 3 to 14 μm, the sky does not become truly transparent again until we reach the optical window of 0.25–1.5 μm. At the other end of the scale the ionospheric sector of the atmosphere acts as a refracting-reflecting agent and prevents radio waves longer than 20m from penetrating to the surface. 20mm to 20m spans 3 decades or about 10 octaves; the radio-astronomer therefore commands a much wider spectrum than his optical counterpart, who deals with a span of only 2.5 octaves — i.e., less than a decade.

The optical astronomer inherently has a better resolution capability than does the radio astronomer, but the latter, by use of such variations as interferometry, has been able to match and now surpass the resolution of optical instruments. Interferometry has been developed to a very high degree of precision and forms the backbone of most modern installations.

The radio effect is the exact counterpart of the optical one

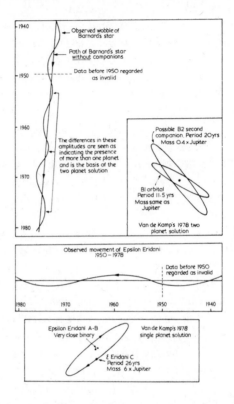

11 The detection of planets of nearby stars. In both of these examples, the results are derived from nearly three decades of work by Dr Peter Van de Kamp. The upper example shows the perturbations of the observed motion across the sky of Barnard's Star owing, according to Van de Kamp, to at least two massive planets; the lower example shows a similar perturbation in the course of Epsilon Eridani, a Sunlike star, owing to its possession of at least one planet, of mass six times that of Jupiter. While further refinements must still be made before Van de Kamp's theses can be regarded as proven, the detection of *any* planets orbiting nearby stars has profound effect on our estimates of the probability of other intelligent lifeforms besides ourselves existing in (relatively) near proximity.

exploited in the Michelson interferometer and used for measuring the actual diameters of the nearest large stars as briefly discussed on page 61. The effect occurs when trains of two or more waves of the same wavelength are combined. The output of the combination depends solely on the phase and amplitudes of the inputs. If the apparatus is carefully matched such that the input amplitudes are equal, then the output is purely dependent on the phase difference between them. This difference can be shown to be proportional to the wavelength used and the physical separation of the two installations.

If the wavelength is denoted by λ and the separation distance by D, then the overall resolution is $r = \dfrac{1.22\lambda}{D}$ radians. This is an exact analogy of the resolution limit of an optical telescope $r = \dfrac{1.22\lambda F}{a}$ where F is the focal length of the telescope and a the aperture of the mirror or object glass.

These formulae clearly state in both cases that the best resolution will be obtained by using the shortest practical wavelength and the largest possible diameter or separation. In early radiotelescopes, the wavelength used was of the order of 1m and the separation about 500m. This gave a resolution of about 8.5 minutes of arc. Gradually techniques improved, using shorter wavelengths with greater spacing between installations. The modern radiotelescope is typified by the system at Cambridge where the 'dishes' may be separated by 5km and operate at a wavelength of (say) 20cm. This gives a resolution of about 10 seconds of arc. If the wavelength were reduced to 2cm, the resolution would be 1″, comparable with an optical telescope.

But the radioastronomer has a further trick up his sleeve, for he can link up with another installation maybe 500km distant; the linking may be done *via* radio or by a telephone land line. *Now* the combined resolution at 1cm is 0.01″ — an extremely difficult if not impossible feat for current optical telescopes. The radioastronomer has not stopped here for installations on opposite sides of the Earth have been linked to provide baselines which are comparable to the diameter of the Earth. This is

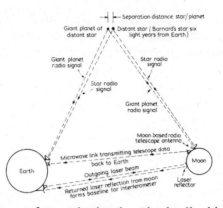

Separation distance star/planet

Giant planet of distant star

Distant star (Barnard's star six light years from Earth)

Giant planet radio signal

Star radio signal

Star radio signal

Giant planet radio signal

Moon based radio telescope antenna

Microwave link transmitting telescope data back to Earth

Earth

Moon

Outgoing laser beam

Returned laser reflection from moon forms baseline for interferometer

Laser reflector

12 Interferograms of unresolved and resolved radio objects. When an object is not fully resolved the interferometer signals, as shown in the upper part of the diagram, produce a 'blur' because they are not in phase. When they are resolved, the signals add up together to produce an ideal central peak, the width of which (at 0.5 height) is the width of the object. If there are two objects then two peaks are obtained, as shown in the lower part of the diagram. Interferograms provide a very powerful means of resolving objects with small angular diameters.

approximately 12,750km and the corresponding resolution is therefore about 0.0004″. Successful links of this kind have been performed between the UK and USSR (Jodrell Bank and the Crimea) and Sweden and the USA (Upsaala and Stanford Universities). These very long baseline interferometry (VLBI) measurements have been used to obtain very fine structure resolution of some of the more distant objects (quasars, Seyfert galaxies, etc.).

It is in projecting still longer baselines that satellite radioastronomy will play a major part. There is little point in placing such objects in close orbit (say 400km) above Earth's surface for these experiments only, for the resolution obtained will be little better than that already achieved. But it is very likely that Large Radio Telescopes (LRTs) will be built and placed in orbit and used for the detection of signals too faint to be traced otherwise (see Plate 20). There is another very long baseline with a solid body at the far end. This is the distance from Earth to

Moon, with a mean value of 384 x 10³km — i.e., 30 times the diameter of the Earth. If this were used as an interferometric baseline we should theoretically be able to measure diameters down to 0.0000133″ at a wavelength of 1cm. Alternatively, a lower resolution is possible at a longer wavelength; for example, the angular resolution at 1m would still be 0.00133″.

If such a system were attempted, the distance between the Earth and the Moon would need to be accurately known. This varies from 356,000km to 407,000km — i.e., by about 51,000km (about 12.5%) — and this must be taken into account. Fortunately this can easily be measured by means of a laser beam aimed from Earth and designed to hit a reflector on the lunar surface. This experiment is already routine and has allowed hyperfine resolution of movements due to tidal action on the surfaces of the Moon and the Earth (these experiments resulted from the ALSEP programme). It was pointed out then that attempts *directly* to link the Earth-Moon baseline with a laser beam would cause difficulties. But there is no reason why the accumulated results of many years of such experiments cannot be used as the basis of an accurate computer programme which takes into account the tidal movements on the surfaces of both the Earth *and* the Moon. These are second-order effects and, for extreme accuracy, it may be necessary to include Earth-Moon perturbation by both Venus and Mars. Again these data can be amassed over years of routine laser sounding. The laser system would then be used only as a corrective monitor of the Earth-Moon perturbation programme. (As a completely incidental by-product, the compilation of a super-accurate determination of such movement would allow the detection of any small asteroid-like bodies which may not show up on photographic plates because of low albedo. They would cause detectable drift perturbations at present lost in 'noise'.)

By the time the Moon becomes the site of a radio observatory we should be able to account for movements of ±0.1m in the mean orbital distance of 384,400km as part of a standard error or allowance in a computer programme.

What purpose could be served by having such very long base-

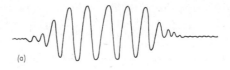

(a)

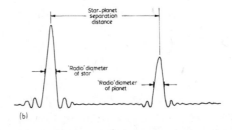

Star-planet
separation
distance

'Radio' diameter
of star

'Radio' diameter
of planet

(b)

13 Radio interferometry using the long baseline between Earth and Moon (about 400,000km) to examine the radio emissions from Barnard's Star and a planet in orbit about it — a diagram which is very much not to scale! The star's emission is in the form of thermal radiation; that of the planet is in the form of synchrotron radiation. Typical wavelengths to check would be 1-10m. Resolution at these wavelengths would be of the order of 0.013 to 0.0013 arc seconds (for 10m and 1m respectively). Such resolutions would easily allow separation of the companions depicted on page 98.

lines? One direct consequence of high resolution at comparatively long radio wavelengths is the ability to detect the presence of planetary companions to nearby stars. The Sun has a surface temperature of 6000 K and emits its peak radiation at 0.5×10^{-6}m. The Earth, with a mean surface temperature of 288 K, emits its peak radiation at 10.4×10^{-6}m.

The giant planet Jupiter emits its peak thermal radiation at 22.2×10^{-6}m from a surface with a temperature of 135 K, but this is dwarfed by radio emission at wavelengths of 1 to 10m caused largely by synchrotron radiation — the electrons spiralling rapidly along the lines of Jupiter's intense magnetic field.

At these wavelengths the thermal radio emission from the Sun is greatly reduced in comparison with its visible radiation and the output from Jupiter is sufficient for a high-resolution high-sensitivity radiotelescope outside the Solar System to separate the two sources. Viewed from another star, the Sun-Jupiter system would appear to be a 'radio binary star'. At present we can detect vague indications of perturbation of some of the nearer stars, up to about 4.5 parsecs away.

The best documented of these is Barnard's Star, a dim red dwarf 2 parsecs distant in the constellation Ophiuchus and possibly accompanied by two planet-like companions. Dr Peter Van de Kamp, ex-Director of Sproul Observatory, Swarthmore College, Pennsylvania, has made a lifetime study of this particular star and is firmly convinced of the reality of the planets. Originally, in 1963, he reported that there was one companion with a mass 1.5 times that of Jupiter orbiting with a period of 26 years and at a distance of 4.5AU — a similar distance to the separation between the Sun and Jupiter. Later, in 1969, he reported a two-planet solution. One planet orbited the star with a 26 year period at a distance of 4.5AU and had a mass 1.6 that of Jupiter; the other orbited at a distance of 2.8AU with a mass 0.8 that of Jupiter.

These results were severely criticized on the grounds of systematic errors in the Sproul instrumentation. As a result of this Van de Kamp published a further two-planet solution, in which the more massive planet has the same mass as Jupiter and orbits in a period of 11.5 years at a distance of 2.7AU while the other revolves at a distance of 4.2AU in a period of 22 years and has a mass 0.4 that of Jupiter. Van de Kamp is not firm about this second body, considering his conclusions tentative.

I have given these results in some detail because they show the problems involved in deducing astrometric perturbation data and extrapolating the results to extrasolar planets. Van de Kamp has examined nearly ten thousand photographic plates taken from 1914 on, has been forced to abandon all Barnard's Star plates taken before 1950 — and is *still* not able to confirm the presence of a planetary companion to the star. And it must be emphasized that the star itself is the second nearest, and is a small star with low mass (about 15% of that of the Sun) so that it will be more easily perturbated by a large planetary companion.

If we could 'see' these two planets of Barnard's Star we would be in a position to confirm their presence, their separation

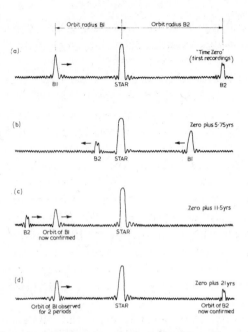

14 Radiotelescope interferograms of stars with radio-emitting giant planets. In this case, the example is of Barnard's Star (see Fig. 13, page 110). If such a high-resolution, high-sensitivity system were to be built it might immediately confirm the presence of companions — as in (a). Further study of the apparent movement (b), (c), (d) would confirm the orbital data and masses. The central star image is shown as a reference, and does not appear to move; if checked against a background reference, it would appear to move.

distance, orbital period and probable mass with much more precision than at present. If we took our Earth-Moon radiotelescope interferometer and directed it at Barnard's Star then, if there are Jupiter-like planets in orbit around it, we would see them as extra sources of radio noise (I assume that Jupiter is typical of a giant-supergiant planet). If they are separated from the central star by 2.7 and 4.5AU respectively, then these sources at maximum separation would be at 1.5 and 2.5″ from the main star radio noise. If the star were examined again (say) one year later, then the radio sources would have noticeably moved: the inner planet would have moved through approximately 35° of circular orbit while the outer one would have moved through

12°. These orbital movements would be noted as angular motion with respect to the parent star central noise source. Observation of the system over 11 years would confirm the inner planetary orbit and observation over a 22-year period would confirm the outer planet's presence. It would also confirm or refute the possible presence of other large planets by the perturbation patterns produced on the observable noise sources. Some analyses of the Sproul perturbation data have suggested that a three-planet or even a five-planet solution is feasible. Van de Kamp is inclined to dismiss *these* multi-planet solutions as 'over computerizing'.

The use of radio VLBI for planet detection in this manner would also clear the question of other nearby stars suspected of having planetary companions. The resolution at 1m of an Earth-Moon VLBI has been quoted as 0.00133″. If the orbital separation of the nearest planet of Barnard's Star is seen as 1.5″ then, provided that the radio emission is detectable out to (say) ten parsecs, the presence of planets of other stars could be distinguished easily. In fact, the system is *noise* limited rather than *resolution* limited; the latter does not fall off until distances exceeding 300 parsecs are involved.

Extraterrestrial Radio Signals
This high degree of resolution would be required if a radio search for extraterrestrial intelligence (SETI) experiment were to produce positive results. It is an unlikely event, but searches at various radio frequencies will continue to be funded on a small scale in the USA and possibly to a greater extent in the USSR. Most of the work has been or will be devoted to scanning the nearer stars at 1,420MHz (the absorption frequency of neutral hydrogen) and 1,660MHz (the absorption frequency of the hydroxyl molecule). The frequency band between these markers has often been referred to as 'the Water Hole'.

As of 1979, Professor John Billingham of NASA Ames Research Centre is proposing a high-resolution search of this complete band. Using a receiver with a 100MHz bandwidth, and a spectral analyser capable of splitting this down to individual channels 1 Hz wide, it is proposed that 400 nearby candidate stars be searched for signs of intelligent signals. The 300m giant radiotelescope at Arecibo is to be used for the initial work, which is mainly seen as an exercise in developing receivers with very high sensitivity and good narrow-band discrimination powers. The actual bands of OH and H have not been searched with high-resolution equipment, and such close examination should produce interesting results from the natural spectrum in terms of hyperfine structure.

Recently Morimoto *et al* of Tokyo University suggested searching around the resonant absorption line of formaldehyde, occurring at a frequency of 4,830MHz. Formaldehyde, because of its structure, is capable only of radio absorption, without any re-emission as is the case with hydrogen and hydroxyl. Any emission line appearing at the formaldehyde frequency can *only* be interpreted as artificial. Indeed, this 'anti-maser' absorption is so pronounced that it causes an actual depression in the 3K microwave background. At the absorption frequency the background temperature is nearer 2K — a useful 30% reduction.

This leads me to suggest a unique set of lines which occur very close together in the same radio frequency band, are related to hydrogen, hydroxyl, and formaldehyde and have low background noise. If transmissions were found there they would unmistakably be artificial.

The frequencies referred to earlier were 1,420MHz and 1,660MHz. If we take the third harmonic of these frequencies, we have 4,260MHz and 4,980MHz respectively. These sit comfortably on either side of 4,830MHz making a unique spectrum which I call 'the Formaldehyde Window'. Let us suppose that a radiotelescope and receiver are carrying out a sky scan at the formaldehyde line. The system would be looking at a gas cloud nebula — let us say the Coal Sack — and measuring the density of formaldehyde concentration by ascertaining the *degree* of absorption. The nebula forms a background about 7.5° wide and about 3.25° deep. Between the nebula and Earth lie 36 solar-type star systems, 11 of them double or binary stars, making a possible grand total of 47 sun-like dwarf stars. Of these, 14 are in the G type solar group, and 4 are G_2–G_5 subclass, almost exact duplicates of the Sun.

If we received an emission signal from this area at this frequency (4,830 MHz) it would most likely be modulated with a narrow-band very-low-data-rate signal — exactly the type expected by Billingham. This signal could very well be coded in such a manner that it provided only two ratios of numbers. Need we be surprised if we found that one ratio expressed the relationship between 4,830 and 4,980 MHz, and the other showed the ratio between 4,830 and 4,260 MHz? Suppose we tuned the system to 4,980 MHz and heard a wide-band signal, heavily modulated, and then tuned to 4,260 MHz and heard just the faint noise of the carrier but no modulation — what would we deduce?

I have asked my professional colleagues this same question — and always received the same answer. 'The signal being heavily modulated at 4,980 MHz and unmodulated at 4,260 MHz can *only* mean that "they" are leaving the channel open at the lower frequency and are expecting you to reply on 4,260 MHz!' There is no other interpretation! It is a simple direct means of conveying information and it is an intelligence test all rolled into one. The system also elegantly avoids the natural maser emission noise which occurs at 1,420 and 1,660 MHz. Nature cannot harmonically multiply, and the background at 'the Formaldehyde Window' is purely the 3 K microwave radiation and nothing else.

If we detected such a signal from somewhere deeper in the Galaxy than the Coal Sack (125 parsecs distant) then we may need the services of the Earth–Moon VLBI to pinpoint and identify the transmitting site.

I sincerely hope that when large orbital radiotelescopes are eventually built (both the US and USSR have designs on paper for such items) they will search the appropriate areas at 'the Formaldehyde Window' as well as 'the Water Hole'.

After all, if you dial a number and get no answer, even if you know there should be somebody there, it's a reasonable assumption that you've dialled the wrong number.

9 Novae and Supernovae

Modern astrophysics theory indicates that a star may 'die' in one of three (possibly four) ways. The great majority that form a galaxy such as our own Milky Way are of a size similar to or smaller than that of the Sun. Such stars are believed to spend the greatest part of their lives on what is called the Main Sequence. They may last for 10 to 15 billion years in this phase, quietly burning hydrogen to helium. When enough nuclear fuel has been used up for over 16% of the mass of the star to be helium, a sudden and major change takes place: the temperature and pressure at the core of the star are now such that the helium, in its turn, starts burning. This extra energy release is almost explosive, but the star is able to contain the surge by very rapid expansion. The increased surface area of the larger globe allows the star to rid itself of the surplus energy by more peaceful radiation, and it then settles down to a different mode of life. The larger surface area is much cooler, and the star is now bloated out to about 200 to 300 times its original diameter. Stars in this condition are aptly termed 'red giants' — 'red' because their visible surface has cooled from 6,000 K to about 3,500 K, and 'giant' because the diameter has increased from about 1.4 million km to about 350 to 400 million km. Thus readjusted, the star settles down to a stable life of several tens of millions of years; by way of comparison the helium 'flash' which initiated this new mode of existence lasted little more than one hundred seconds.

It is believed that this change will happen to our Sun in about 4 to 5 billion years' time. The theories which indicate this evolutionary future suggest also that the Sun is approximately 5.5 billion years old, and is therefore halfway through its Main Sequence phase.

During the next 5 billion years the Sun will slowly grow brighter as its internal helium core increases in size. At present this core is very small, since the helium content of the Sun is believed to be only about 8% of the total mass: the balance is 90%

hydrogen, with iron, copper, nitrogen and oxygen — the elements so familiar to us on Earth — making up less than 2%.

As the size of the core increases so does its surface area — think of blowing up a balloon. If the core diameter were to double, the surface area would be increased four times. All things being equal, the brightness *should* also increase in the same ratio; in fact, the brightness increases by a much larger factor than the implied simple square law.

In fact, the Sun will enter the helium-burning phase and evolve into a red giant long before the present helium core doubles in diameter — in fact, when it reaches about 1.26 of its present diameter. Under red-giant conditions the inner planets Mercury, Venus, Earth and Mars will vanish leaving the outer planets circling what has been described as a 'red-hot vacuum'.

Even the outer planets may not last, for in the red-giant phase a star begins to lose appreciable mass by what is called 'stellar wind'. The Solar Wind emanating from the Sun at the moment is very weak (about five protons in each cubic centimetre of space in the vicinity of Earth). Consequently its effect in terms of physically slowing down any of the planets in their orbits is negligible. In the red-giant phase, however, the stellar wind is increased by a factor of several thousand and it is quite feasible that the combined effects of physical drag and the magneto-electric interaction between a planet's magnetic field and the plasma of the stellar wind would begin to affect that planet — even one as massive as Jupiter. I therefore consider that, once a Sun-like star comes off the Main Sequence and becomes a red giant, it probably loses *all* of its attendant planets, the innermost ones more or less immediately, the outer ones a little while later.

The well known red giant Betelgeuse (α Orionis) has recently been very accurately measured using the techniques of speckle interferometry. Calculations had shown that the diameter of Betelgeuse should be about 400 times that of the Sun, and measurements made by Michelson between 1924 and 1930 using the 2.5m Mount Wilson mirror were in general agreement. The most recent measurements, of 1974-75, tend to confirm this diameter for the *main* body of the star, but show also that the outer layers extend much further than had previously been believed. Furthermore, the whole of this outer surface is a seething turmoil of convection currents, rather as if it were boiling.

Although a red giant may have a mean surface temperature of about 3,000 K, in different areas the actual temperature may lie anywhere from 2,000 K to 10,000 K or even higher. Since the diameter of Betelgeuse is just less than the orbital diameter of Pluto, virtually the entire Solar System would be lost in such a body.

Novae—The Flash without a Bang

Rather like an audience watching a play for the first time, one may well ask: 'What happens next?' Different answers have been offered, particularly in the last twenty years. For stars such as the Sun, it was thought that perhaps the red giant initially experienced a 'flare up' or explosion called a nova (Latin *nova stella* or 'new star'). This was thought to be due to sudden mixing of the relatively cool outer hydrogen shell with some of the super-heated helium rising by convection from the core: the effect would be akin to hosing a fire with petrol.

Novae have been seen by both ancient and modern astronomers; the latter have been able to record the rise in brilliance as the event developed. Photographs have shown a bright shell of hot gas surrounding the star and travelling away from it at about 1,600km per second (approximately 0.5% of the speed of light). Calculations have shown that the star has lost about 10^{-4} (i.e., about one ten-thousandth) of its mass, mostly in the form of hydrogen — this has been confirmed by spectral analysis. The spectral lines show traces of iron, calcium, carbon, oxygen, helium and other elements similar to those found in the Sun.

But, as we have seen, the stars that underwent these violent changes seemed to lose very little mass in the process, and it was natural to assume that it could happen more than once. Several novae have now been observed to have repeated flare-ups, thus again seemingly confirming the intermixing convection theory.

18 A visualization of a possible future rendezvous between the NASA Shuttle and the Soviet Kosmolyot (in the background). The Soviet craft is believed to be rather smaller than the Shuttle. (*David Warmington.*) See page 49.

19 A visualization of NASA's proposed Mars Rover, intended to land on the planet during 1984 and, during a Martian year, to traverse some 100km of territory. It carries various equipments for scientific survey of the planet; most interesting, however, is the fact that within the meaning of the act it would be intelligent, able to make its own decisions and to act independent of detailed instructions from Earth. (*Courtesy NASA.*) See page 92.

Then, in the late 1960s, an alternative theory was presented which rapidly overhauled the 'petrol bonfire'. This theory was based on the observation that all the novae close enough to allow the actual stars to be seen were binary systems. What was even more striking was that one of the components was invariably a red giant, while the other was a very hot-surfaced body and most probably a white dwarf, although X-ray astronomy has shown X-ray novae which may be linked with neutron stars — in either case, the object has a surface temperature of 30,000K or more. Calculation showed that the binary components were close enough for the large red giant to be quite near to or just within the Roche Limit of the very hot companion.

We have seen earlier that red giants are literally unstable quaking jellies of distended hydrogen so, if we now imagine another star lurking nearby, this latter could well snatch off a fraction of the red giant's bulging outer envelope and suck it down onto its own hot surface. This is akin to dropping a giant napalm bomb on a bonfire rather than merely hosing it with petrol! But there is another effect which gives a major impulse to the energy release, the kinetic energy picked up by the hydrogen as it 'rollercoasts' down the steep gravitational well of the companion, gaining in speed and momentum until it crashes onto (and into) the superhot surface.

This sudden energy release raises the temperature to well over the thermonuclear ignition point for hydrogen. It promptly flashes into helium and other elements (depending on temperature and pressure). The resultant nuclear reactions blow the unused hydrogen, brilliantly glowing, into space.

A 'Quiet' Death

Because it is a single star and not a member of a binary system, there are good grounds for thinking that a different fate awaits the Sun once it has evolved to the red-giant phase. There are differing schools of thought, but the prevailing opinion is that helium-burning will be maintained at the core while the outer-layer hydrogen-helium reaction continues in parallel. A star in this condition without a companion to disturb it will merely grow larger as the core and skin areas increase with progressive burning. Eventually the outer layers become so tenuous that they drift off into space, for the star's gravitational field can no longer control them. If this takes place uniformly, then the dying star may generate a planetary nebula.

Several of these objects have been discovered but no satisfactory theory has so far fully explained the observed characteristics: a small, very hot central star is surrounded by one or more expanding shells of hydrogen. The shells glow with the characterised red of H_α, the energy for this being derived from the central star, which appears to be of high density and possibly just one step removed from being a white dwarf. On the basis of this theory, the star finally 'dies'; i.e., it is no longer capable of generating energy by nuclear processes. Although the hydrogen and helium fuels have not been completely used, the star has insufficient mass to create the gravitational pressure required to compress the core material to the next stage of energy-release.

The star therefore 'stops', or turns off, with a dense compressed core of carbon and 'light' elements surrounded by a thick skin of helium and some traces of hydrogen. The core material is collapsed to the point where the central atomic nuclei are almost (but not quite) in contact with each other. In this condition a star is said to be degenerate. It cannot produce energy except by gravitational contraction. It therefore progressively shrinks and cools until finally it is no longer emitting light. Since the radiating surface area is very small this takes a long time — the known Universe does not appear to be old enough for any stars yet to have reached the 'black dwarf' phase.

But stars which may be at the very beginning of this long decline and decay *have* been observed. The visible surface is white-hot, and the star itself is extremely small: hence it is called a 'white dwarf'. The first discovered and best known example of such an object is the companion to Sirius (officially called Sirius B but generally known as the Pup, for obvious reasons). It is only 40,000km in diameter but its mass is 90% that of the Sun, which means that it has a density over 1,000 times that of water: a sugar cube of Sirius B material would weigh about one tonne. The surface temperature of Sirius B was originally thought to be about 8,000K, but accurate satellite measurements in the far ultraviolet have shown that it is about 30,000 K.

Currently, therefore, we expect the Sun to go through the red-giant phase, probably entirely destroy its attendant planets, quietly expire in a planetary nebula, and then become a solitary white dwarf, bankrupt of energy and companion planets. 'Sans eyes, sans teeth, sans everything' is an aptly fitting epitaph for such a 'dead' star.

Death of a King

In 1937, two US astronomers, Zwicky and Baade, during examination of photographs of nova-like star explosions in other galaxies, noticed that some events seemed to be extremely violent. Baade then recalled the accounts by earlier astronomers of the appearances of super-brilliant objects in our own Galaxy. These were so bright that they were visible in broad daylight for several weeks, and were visible in the night skies for some years before fading from sight. 'New stars' were recorded in AD 185, 1006, 1054, 1181, 1572 and 1604.

The event of 1572 was minutely detailed in Europe by Tycho Brahe as regards position and brightness when compared with other stars; his records were independently confirmed by Thomas Digges — mathematician and astronomer of Cambridge. Using Tycho's results Baade was able to construct a light curve of the 1572 outburst. He showed that it closely resembled the light curves of the outbursts observed in his photographs of other galaxies, and he classed the 1572 event as a 'Type I' supernova (the classification and meaning of which is discussed later). Baade then repeated the light curve exercise for the star of 1604 using the figures recorded by Kepler, Tycho's understudy and successor as Imperial Mathematician at Prague — and, incidentally, one of the most important figures in the history of astronomy!

The close similarity of these curves convinced Baade and Zwicky that there *were* rare events which occurred in the Milky Way as well as in other galaxies. Although they appeared as 'new' stars, they were in reality explosive outbursts of such violence that they exceeded the energy release of an ordinary nova by a factor of about 10^5. Since a nova produced for a short period about 10^6 times the normal output of the Sun, this meant that, during a period of a few months when such a super explosion was at its peak, one star alone produced almost as much light as all the rest of the stars in its galaxy! To these immense outbursts Zwicky and Baade gave the name 'supernovae', for quite clearly they were over and above the ordinary nova outburst.

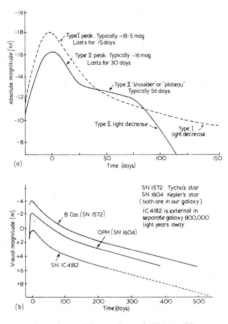

15 The difference between Type I and Type II supernovae curves (*above*), showing the greater violence of the former. In the lower part of the diagram are shown the curves that confirmed the existence of supernovae.

Such explosions also seemed to represent the end of a star, for the mass loss required to produce the outburst meant that it probably happened only once and could not be repeated, as with novae.

A glowing patch of gas was found near the area reported by the Chinese for the supernova of AD 1054. Because of its shape, the first man to see it clearly (Lord Rosse) named it the Crab Nebula. In 1968 an object with remarkable properties was discovered deep in the heart of the nebula. But of an ordinary star there was no trace — it had gone out!

When Zwicky and Baade announced their findings, there were initial difficulties in explaining why a star should explode in a spectacular way. Theoretical work by astrophysicists in the early 1950s — especially Hoyle, Geoffrey and Margaret Burbidge in Europe, and Fowler and Salpeter in the USA — showed that stars larger than the Sun could synthesize all the elements up to and including iron, because the larger mass involved could generate the required temperatures and pressures to perform the necessary nuclear transmutations. This early work gave an accurate insight into the temperatures and pressures expected at the core of a star and in the various shells or zones around the core.

Thus, at the very centre of a star shortly destined to become a supernova, there may be an ultradense core of iron and the 'iron elements' (nickel, chromium, manganese, and cobalt); the temperature may be above 2,000 *million* K and the pressure about $7 \times 10^{19} \text{kg/m}^2$. Surrounding this is a shell at a slightly lower temperature where silicon is 'burned' to iron; this in turn is enclosed by another shell burning oxygen and magnesium to silicon. Other shells are in the intermediate phases of burning helium to carbon and carbon to oxygen and neon.

This theory therefore envisages a complete hierarchy of nuclear fuels, each being consumed to provide energy at progressively higher temperatures and pressures. This nuclear merry-go-round stops drastically when iron is reached. To progress to elements of greater complexity *demands* energy as opposed to producing it. Iron is the crossover point in the Periodic Table of the Elements, whereby more complex elements yield energy by splitting up (fission), as typified by uranium and plutonium, while less complex elements such as the ones we have been concerned with yield energy by melding together (fusion).

The star is now at crisis point, for the central energy generation has stopped and there is no radiation pressure to hold up the mass of the still-burning overlying shells. The result is a swift collapse followed by a catastrophic outburst, for in collapsing the layers gain gravitational energy, become rapidly superheated and explode. The star simply cannot adjust itself, as it did when helium-burning started. The energy release is too great, and the outer layers are torn off. Relieved of the gravitational restraint of the shells above it, the core breaks its bonds and pours into space, leaving a small remnant that probably becomes a white dwarf — now modified to a neutron star.

This theory was modified following the discovery of a pulsar at the centre of the Crab Nebula. It is now generally conceded that pulsars are neutron stars, stars made of material that is far more dense than even the core of a white dwarf. A typical white dwarf may have a density of 10^6 grams per cubic centimetre, whereas a neutron star has a density about 10^{14} g/cm^3 — that is, 100 million times greater. *If* our Sun were a white dwarf it would have a diameter of 30,000km, but if it were a neutron star it would have a diameter of barely 16km, but with the same mass. The mass of the Sun in a body the size of the Isle of Wight or Long Island is a bizarre concept but the facts support the hypothesis.

Later theories concerning supernova evolution invoke evolved binary star systems in much the same manner as for an ordinary nova. But first it is necessary to distinguish two main classes of supernova.

Type I and Type II Supernovae
Baade distinguished between two main types of supernova by comparing the peak light output achieved and the length of time the supernova would be above naked-eye visibility (generally assumed to be about magnitude +6). A Type I supernova has an absolute magnitude of −19 at maximum, and, after an initial drop of about 3 magnitudes in about 25 days, the curve drops exponentially, with the light output approximately halving every 20-30 days. It declines by about 4 magnitudes after 100 days, approximately 7 magnitudes further after 300 days

and a further 10 magnitudes after 3 years.

A Type II supernova behaves differently, and displays a pronounced 'shoulder' after the maximum. This peak output is itself slightly less than for a Type I outburst, being about magnitude −18. The hump of the peak lasts for about 20 days, during which time the light output has dropped about 3 magnitudes. The 'shoulder' then appears, as the light curve levels off: during the next 50 to 60 days the output drops by only 1 magnitude. Then the curve falls away steeply and drops by 6 to 7 magnitudes in the next 50 days. These latter figures are just an outline, for only a few Type II supernovae have been observed (in other galaxies) and their behaviour is somewhat individualistic. However, the curves are quite distinctive and cannot be confused with that of an ordinary nova: the nova curve has been included in our diagram for comparison purposes only, and its lower peak magnitude is immediately obvious. Nova peaks also last for a shorter time; the peak of a 'fast' nova may last only 5 to 10 days, after which it will slowly sink back, reaching its original low state after about 1,000 to 2,000 days.

As stated earlier, Baade plotted the brightness curves of the 'new stars' observed by Brahe in 1572 and Kepler in 1604. These were both Type I supernovae; the curves are ample evidence of the accuracy of the medieval observations — and a perpetual tribute to the observers themselves.

Identification and classification of ancient supernovae has been carried a stage further by Stephenson and Clarke, who have heavily researched Far Eastern records. They have been able to classify the supernovae of AD 185, 1006 and 1181 as Type I. This has been possible because the remnants of these outbursts have been identified and their distances estimated. These remnants are normally radio objects, although optical remains have also been identified for the AD 185 and 1006 items.

Identification of these objects from their optical remains is developing into a skilled art. The filamentary wisps of glowing gas may eventually develop a characteristic shape which allows the observer to tentatively identify a Type I or Type II event — even though the outburst may have occurred several thousand years ago. The age record for optically identified supernova remnants presently stands at 300 thousand years! This prehistoric remnant was found in the constellation of Cygnus by Gull, Kirschner and Parker. Using an electronic image intensifier and the superb 4m telescope at Kitt Peak they were able to photograph the thin skin of gas, still fleeing from the centre of the explosion at 160,000km per hour and now having the form of a huge egg-shaped shell nearly 70 parsecs in diameter! The distance from here to the remnant (discovery of which was announced in 1977) is about 1,200 parsecs. From the shape of the shell it is highly likely that the supernova was of Type I. Cygnus is a Summer/Autumn constellation and during these seasons is well visible during the night in the northern hemisphere; our prehistoric ancestors would have seen an object which rivalled the quarter Moon in brightness! It probably cast flickering shadows as the light twinkled in the sky.

The differentiation of Type I and Type II curves for supernovae shows that there are probably at least two distinct mechanisms at work. The general opinion is that Type I outbursts are from relatively small stars of up to about three to four solar masses, while Type II are from stars larger than this. Note that there is also a *lower* limit of about 2.5 solar masses below which the star will not explode, because it cannot raise the necessary temperatures and pressures to achieve iron synthesis and collapse of the central regions.

We have not been able to observe and identify any stars *before* they explode as supernovae, and so all the pre-outburst work is theoretical nuclear physics. However, the spectra of several supernovae in other galaxies have been examined and the presence of iron and other minerals positively identified. We can therefore hypothesize the conditions prevailing just before the explosion and visualize the appearance of the star.

In a Type I supernova we have very tight mass limits (2.5 to 4 solar masses), and we are dealing with immensely high pressures and temperatures. The star could therefore be small and hot, and one may guesstimate the diameter as being about 15% of that of the Sun, but with an external temperature of about

50,000 K. The star would therefore be small and blue with possibly very little visible light being emitted, the bulk of the radiation being in the form of neutrinos, X-rays and ultraviolet light.

Immediately after the explosion, the bulk of the relatively thin surrounding skin would be ripped apart, leaving the super-hot iron core free to expand into space to give the observed leap to the peak light output of magnitude −19 or so. As the core rushes out, the falling temperature is to some extent offset by the increasing diameter of the expanding gases and the total light output is relatively unchanged for the first 20-25 days. Gradually, however, the temperature falls off faster than the expansion rate and light output falls off exponentially.

Conditions in the Type II event are slightly different. Here the star is much larger and more massive. It is therefore possible that the outer layers may be somewhat cooler (say 10,000 K-20,000 K). The explosion, when it occurs, does so deep inside the star and as a result the effect is muffled by the outer layers. The result is that the light output from the brilliant central core is reduced; moreover, the high opacity of the thick expanding shell causes a steep drop in the initial 'flash'. However, the intense radiation impinging on the blanketing shell causes secondary nuclear reactions which sustain the output at a more or less constant level, thus producing the characteristic 'shoulder' of a Type II event. Finally, the shell expands under the sheer pressure of the shock wave produced by the outburst and the light curve dies away in the observed manner. Type II explosions are not as common as Type I, and all those observed so far have been in other galaxies. This is not altogether surprising, for very massive stars comprise only a small fraction of the total population.

Chandrasekhar's Limit

A discussion on the formation and classification of supernovae would be incomplete without mention of Chandrasekhar's Limit, so called after the Indian mathematician/astrophysicist who first pointed out that a white dwarf could not have a mass greater than 1.44 times that of the Sun: above this critical limit, the star would collapse into yet denser materials. The temperature is thus raised by gravitational compression and the star begins to 'burn' the helium to other materials higher up the nuclear scale. It is believed that this could be the first step towards a supernova 'doom'. Of course, the concept of a white dwarf consisting entirely of helium is purely theoretical, but it is an adequate yardstick for judging the reasons why there are no supermassive white dwarfs or (for that matter) no super-lightweight neutron stars.

Chandrasekhar's Limit is sometimes interpreted too literally, and it must be emphasized that it is a 'guide' limit, not an absolute one. As an example, it is quite feasible to consider that a star of 1.45, 1.5 or even 2 solar masses may still be able to dispose of the excess mass, either by expelling the matter as stellar wind or as a series of planetary nebula shells. A star can even hand over excess mass to a nearby white dwarf companion: a star of (say) 2 solar masses may be able to hand over the excess mass to a white dwarf of only 0.5 solar mass; there could be an energetic series of nova outbursts, but in the end both stars would 'die' peacefully as white dwarfs of about 1 to 1.2 solar masses each — the surplus mass being blown off by the nova explosions or as stellar wind particles.

However, if the companion is a massive white dwarf, itself already close to Chandrasekhar's Limit, and if too much mass is handed over, it is the *companion* that vanishes in the fiery flames of a supernova! This is given as a possible explanation for one or two Type I supernovae which seem to be more violent than normal. The Crab Nebula may be just such a system.

The effect of Chandrasekhar's Limit on stars of increasing mass is shown in the table below. In the more massive stars, the limit is (as expected) a very small fraction of the total mass and, as a result, the outer layers of massive stars do not become heavily involved in supernovae reactions. Since these latter stars are thought to be the seat of Type II supernovae, the outer layers, as we have seen, tend to muffle the explosion at the core and so give the modified light curve characteristic of such an event.

VALUES OF CHANDRASEKHAR MASS LIMITS COMPARED WITH TOTAL STAR MASS AND PROBABLE DESTINY OF STAR

Percentage mass at Chandrasekhar Limit of helium	Total Mass of Star (Sun = 1)
100	1.44^a
90	1.6^a
80	1.8^a
70	2.0^a
60	2.4^{b1}
50	2.88^b
40	3.6^b
30	4.8^{c1}
20	7.2^c
10	14.4^c

Key to Eventual Destiny

a   Planetary nebula surrounding a white dwarf.
b   Type I supernova.
c   Type II supernova.
b1 Borderline case where the star may evolve as supernova *or* white dwarf depending on circumstances.
c1 Borderline case where the star may evolve as Type I *or* Type II supernova depending on circumstances.

The table shows another interesting feature, namely the fairly narrow mass limits over which a star may become a Type I supernova. The distribution of star masses in the Galaxy is directly analogous to the boulders, rocks, pebbles and grits of the seashore. For every boulder there may be tens of rocks, thousands of pebbles and millions of grits; i.e., the smaller the size, the larger the population. So it is with the star population of the Milky Way. Only a few stars exist with a mass 100 times that of the Sun, while, right at the other extreme, stars with a mass equal to or less than that of the Sun number about 120 billion! (This assumes that the Galaxy contains about 180 billion stars, and that the dwarf stars — of which the Sun is one — constitute over 60% of the total population.)

So the greatest percentage of the galactic star population is doomed to a 'planetary nebula — white dwarf' death. Planetary nebulae are not particularly bright and white dwarfs even less so, and consequently we see comparatively few of them (about 1,000). From this small sample it is estimated that some 60,000 planetary nebulae exist in the Galaxy. But this ring-like phase is temporary, whereas a white dwarf cools down so slowly it may be regarded as permanent, and on *this* basis there may be *1 to 5 billion* such objects scattered through the Galaxy, each one a feeble glow worm, the dying ember of a once brilliant sun.

We can estimate roughly the total number of supernovae remnants that should be in our Galaxy from the rate at which they are observed to occur in our own and other galaxies, and from the age of our Galaxy. The rate at which supernovae occur in the Galaxy is roughly 1 per 150 years. (Some authorities quote 300 years, others as low as 50 years — I have chosen a median value.) The age of the Galaxy is generally accepted as about 15 billion years and, if we divide this by 150, we obtain a figure of 100 million supernovae having exploded since the formation of our Galaxy. They are therefore rare events (the most optimistic figures indicate a population of about 500 million), and yet they are the source of all the metals and other complex elements which are incorporated into our body chemistry and which are now essential to our civilization and technical development.

The 'Death' that Creates Life

It is therefore fortunate that a supernova *explodes* with such violence, for it ensures the scattering of life-producing elements throughout the Galaxy. If a star 'died' by just quietly collapsing, or huddled the layers of formed elements close to itself like a white dwarf, then life — our life — would be nonexistent. We need iron, calcium, sodium, phosphorus, carbon, nitrogen and chlorine in quantity in order to stay alive, and traces of magnesium, sulphur, potassium, iodine, zinc, manganese and many other elements in order to remain in good health. The white dwarf star synthesizes some of these items but keeps them to itself. The supernova literally sacrifices itself in order to give life, for only in the star's late stages are the heavy elements formed, and in the final explosion there is sufficient temperature and pressure to provide the energy *input* to form the elements over and above iron. The gold, silver, and platinum jewellery we wear and the copper electric cables we use were formed in the furnace and welded in the shockwave of a massive star's death throes.

But that is not all! The outgoing shockwave not only spreads over a distance of parsecs, it carries within itself the formed elements and pushes *in front of itself* enormous quantities of neutral hydrogen. Originally a passive bystander, this material is hurried along until it meets (say) a large cloud of dusty vapour and gas (water vapour, formaldehyde, alcohol) that has collected together in space and is 'wondering' if it should collapse and form a star. The shockwave and the neutral hydrogen hit this dithering mass with all the grace of a punch in the solar plexus (no pun intended). The cloud responds in exactly the same way as a human being — it folds up. The shockwave energy is absorbed and the cloud starts forming into a flat plate, the impact having imparted some degree of spin. The flat plate finally collapses in a specific manner and a proto-sun is born — complete with a placental planetary system tied to it.

This theory of cloud collapse being brought about by a supernova shockwave is very recent and arises from three areas of evidence:

a) The shockwave of the Cygnus prehistoric supernova found by Gull, Kirschner and Parker carries with it a much larger shell of neutral hydrogen.

b) A second object in Cygnus called MWC 349 has been resolved by a team from Steward Observatory, Arizona. They report that the object is only 1,000 years old and is a giant disk star about ten times the diameter and thirty times the mass of our Sun. The star is now definitely collapsing to form a proto solar system.

c) A Solar System abundance of magnesium-26 could arise only from the decay of its progenitor aluminium-26 — which can *only* be formed in a supernova, according to Cameron and Truran in an article published in 1977 in the astrophysics magazine *Icarus*. They do not think that a supernova fortuitously occurred just before the proto solar system cloud collapsed: they believe that the supernova shockwave *actually caused* the collapse in the manner I have described. Cameron and Truran think that perhaps 2% of the very heavy elements came from the supernova shockwave and *not* from the condensing cloud.

The consequences of this new theory are possibly not yet fully appreciated, for *if* there is further evidence that supernovae *are* essential in order to initiate planetary proto system collapse then certain definite constraints may possibly be placed on the number of life-bearing planetary systems in our Galaxy. Earlier we calculated that about 100 million supernovae may have exploded since our Galaxy was formed. The Solar System originated about 5 billion years ago, which means that approximately 70 million supernovae exploded before that and about 30 million have occurred since.

If I may make the assumption that on average only one planetary system emerges for each supernova explosion, then, by definition, only 100 million planetary systems exist in the Galaxy and of *that* total some 30 million were (or are being) formed after our own Solar System was born.

But the 'solar type' star population is some 120 billion stars! The inference is that either supernovae are much more frequent or perhaps one supernova can collapse many clouds into stars — or perhaps a collapsing cloud breaks into more than one star. Perhaps there is *another* mechanism which also causes stars to collapse — after all the very first stars were born without the induced labour of a supernova to aid them!

If we take our optimistic figure of 500 million supernovae and compare it with 120 billion solar-type stars, then each outburst would be required to generate 240 stars. This may be reduced by allowing for the original 'old' stars which were formed without supernova aid; these can be identified by being 'metal-deficient'; i.e., they formed *before* supernovae had time to produce the heavy metallic elements that are found in 'metal-rich' stars. Metal-deficient stars are also known as Population II stars and are the Galactic Methusalahs, while metal-rich stars, known as Population I, are comparative newcomers. Our own Sun is a metal-rich Population I star. If I again make a simple assumption, that the Galaxy is made up of 50% Population I and 50% Population II, then each supernova is still required to produce about 120 stars.

The clouds from which the stars condense are fairly massive objects in their own right, and may be about 30 to 200 times as massive as the Sun. However, we cannot guarantee that a cloud of 100 solar masses will automatically collapse to form 100 stars. Most likely it will produce maybe one or two and allow the rest of the gas to escape; and this is the main factor that suggests that most clouds collapse spontaneously, *without* supernova aid. They may produce planetary systems but there would be a vital difference. These systems could be sterile.

We therefore come back to a figure of one proto solar system being formed per supernova explosion, or about 100 million in our Galaxy since the Universe came into being, among which is numbered our Solar System. Although other stars may have condensed by a different process, it is just possible that the *supernova-induced condensation added the right catalytic elements to the proto-life molecules in the gas cloud.* These in turn formed the macromolecules that eventually became the basis of living cells floating in the 'primeval soup' in the oceans on Earth's surface.

We may therefore belong to a privileged group of stars that condensed by a slightly unusual physical process — and each of the other stars in this group may have some form of life-bearing planet in attendance. We will be more concerned with the 70 million planetary systems formed before the Solar System than with the 30 million that came afterwards, for the former may be sending signals to possible other and younger communities — such as that on an insignificant planet orbiting a minor yellow dwarf star called the Sun.

The Crab—A Special Supernova?

We now know that the Crab Nebula is all that remains of a star that ended its career in a supernova explosion, the light of which was seen and recorded by Chinese, Japanese and Arabian astronomers.

Stevenson and Clarke consider that the Crab Nebula supernova was possibly neither Type I nor Type II. Minkowski considered it could be regarded as a special Type I outburst. The difficulty arises because we know of nothing with which we can compare either the apparent characteristics of the original explosion or the present remnant. Even the pulsar in the Crab behaves differently from the two others that have been optically identified to date.

This has led R.K. Kochhar of Gottingen University, writing in *Nature,* December 1978, to conclude that the Crab is actually the result of a *double* outburst. The star is postulated as originally being a binary system. The larger member exploded three million years ago in the constellation of Gemini. The remnant and the companion star were then ejected rapidly from the area at about 0.4% of the speed of light until they arrived at the present site in Taurus. At this time the companion star exploded and the resultant bombardment of the first remnant affected it to such an extent that it joined in. Although it did not truly explode, the companion added to and modified the radiation output. Two pulsars may possibly exist in the Crab Nebula; one has been optically identified while the other has not, but some suspicions might point to the second star placed just above and to the left of the optically pulsing lower star in the illustration opposite.

Using larger telescopes and very high-gain photomultipliers together with sophisticated computer-processing it may be possible to find optical pulses corresponding to the radio 'beeps' from the second pulsar. This might then confirm that the Crab

20 An artist's preconstruction of NASA's LOFT baseline concept. (*Courtesy NASA.*) See page 112.

21 Project SETI (Search for Extraterrestrial Intelligence) deals with the question of whether or not advanced technological civilizations exist in our Galaxy, and if we can receive radio communications from them. One of the tools planned to help in such a search is the Cyclops array, an array of a large number of radiotelescopes acting in concord; while such a tool would be of inestimable value if erected on the Earth, its value would be multiplied many times were

Nebula was indeed a (so far as we know) rare event — a double or binary supernova. The recently discovered Taylor Hulse binary confirms the reality of such systems. See page 172.

Nearby Supernova Candidates

When the sheer scale of a supernova outburst is realized, and the type of star destined for this end has been described, it is reasonable to ask if there are any nearby stars that *will* go supernova, and what will be the consequences for Earth and the life upon it when they do?

The only really nearby star that approaches the mass limits for a supernova explosion is Sirius. This is the brightest star in our sky (magnitude -1.4) and is the nearest star with a mass appreciably larger (2.35 times) than that of the Sun: it is of 1.76 times solar diameter and 26 times as luminous; it is classified as an A_0 star, and has a surface temperature of 8,000 K — about 30% higher than that of the Sun. This higher temperature gives it the characteristic 'electric blue' appearance which, together with its brightness, makes it possibly the most beautiful naked-eye object in the night sky.

This brightness arises not only from its absolute brightness but also because it is the sixth nearest star system to us, about 2.7 parsecs away. Cosmically speaking, Sirius is on our back doorstep. Being larger, brighter and more massive than the Sun it consumes its nuclear fuel at a vastly greater rate. The energy supply for the Sun will last another 5 billion years or so, but that for Sirius will last only about a tenth of that time, or about 500 million years.

Assuming that no disaster has overtaken the human race meanwhile, what will happen then? During its Main Sequence period Sirius A (the larger component) will swell and grow to a red giant. Referring to the table on page 135 we see that its critical mass is close to Chandrasekhar's Limit — it would be a borderline case. If this question had been asked twenty years ago most authorities would have suggested that a star of this mass *would* go supernova. Now the majority would be inclined to think that it would lose an appreciable part of its mass by stellar wind during the giant phase; this would bring it below Chandrasekhar's Limit.

So far, all is straightforward stellar theory, but in the above reasoning we have neglected one important fact. Sirius A is not alone: it has a companion, Sirius B, which is a white dwarf with a mass close to that of our Sun. This will have a profound and complex effect on the ultimate behaviour of the main component. The present situation is quiet, with relatively little mass interaction between the two stars — with the exception of their mutual gravitational attraction. As the stellar wind from the main component increases it will give rise to 'drag' and tend to slow down the white dwarf companion. If this process were to go far enough, the two stars would gradually move closer together and, if sufficiently close, the companion would start stripping the outer hydrogen layers from the now red-giant main component. This process would not be continuous in the early stages because its orbit is elliptical. There would be times of close approach when the action would be violent and vicious and conversely times of recession when the action would be almost non-existent.

It will be recalled that these are the exact conditions for producing nova outbursts — and that is the most likely fate for Sirius. At regular periods it will blaze up as the hydrogen fuel from the outer envelope careers down the gravitational precipice of the companion and crashes onto its super-hot surface. To us here on Earth the effect will be utterly spectacular! Prior to this we will have noticed that Sirius is changing in both colour and brightness. A brilliant helium flash (visible in daylight should it reach us then) will be followed by a swift increase in brightness from magnitude –1.4 to about magnitude –4 to –5 (an increase in brightness of about 10 to 25 times). In this phase it will be slightly brighter than Venus at maximum, but a brilliant orange red. The term red giant will take on a real meaning!

Then, when the first nova outburst occurs, the brilliance will leap in a few hours to magnitude –13.5. This is about six times brighter than the full Moon, which means that most of the stars normally visible in the night sky will vanish whenever Sirius rises above the horizon. The nova will also be visible in full daylight and astronomers (if any) will have a field day observing a nova in full spate at such close range. Fortunately, the outburst will not be sufficiently close to allow any harmful radiation (X-rays, etc.) to approach danger levels at Earth's surface.

In summary, therefore, it is likely that Sirius will one day become a nova; it will outshine all the other stars and be easily visible in broad daylight — but it will not become a supernova.

If Sirius could become a supernova, however, things would be on a vastly larger and much more dangerous scale. It would almost certainly be a Type I supernova and would therefore reach a peak absolute magnitude of –19.6; this would translate to a visual magnitude of –21.0, a level which is about 70 million times the present luminosity, and only about 100 times less than midday sunlight. There would virtually be perpetual daylight whenever Sirius was above the horizon. But much more serious would be the effects of gamma- and X-ray radiation which, in and above the atmosphere, would definitely rise to lethal levels. Most serious of all would be the shockwave which, arriving about six centuries after the outburst, would still be saturated with 'hot' radioactive particles. Details of the effects expected from the explosion of a supernova 100 parsecs away have been quoted by Shklovskii and Rudermann. They are indeed drastic:

a) The cosmic radiation level would rise by 100 times, thus producing a background radiation rise of about thirty-fold. This could be expected to be fatal for some species, for the shockwave would take several centuries to pass through the Solar System.

b) The gamma- and X-ray levels would produce quantities of nitrogen oxides which would completely destroy the ozone layers. At best a few per cent of the ozone would remain.

c) As a consequence of *b* the ultraviolet barrier in the atmosphere would be broken and the Earth saturated with uv — lethal to more complex lifeforms living on land. Marine and aquatic creatures would probably survive.

This is expected for a supernova *100 parsecs away*. If Sirius were to detonate as a supernova the effects from an object less than 3 parsecs away would be almost 1,500 times worse. This would almost certainly destroy all but the very simplest lifeforms: it could be regarded as the end of the world.

Fortunately there are no such nearby supernova candidates, but these results have led Shklovskii and Rudermann and others to speculate that more distant supernovae may have had worldwide climatological effects in prehistoric times, including such events as ice ages and the reversal of the Earth's magnetic field.

Space Experiments and Supernovae
If a supernova in our own Galaxy and within easily observable range of Earth were to occur now we could bring a formidable battery of instruments to bear on the star concerned. We are possibly now in the position to notice the background rise of neutrinos which would occur as a star approached the iron-generation point and collapsed (see page 156). Within two decades we may be able to determine the direction from which the neutrinos were coming to within (hopefully) a few degrees. This would enable different types of space telescope to be oriented in approximately the right direction. Allowing for a certain degree of optimism the operation may possibly proceed as follows:

a) Ground-based neutrino telescopes detect a definite rise in the background count. After checking that this is *not* a solar-activity anomaly, they attempt to detect from which direction these ghost-like particles come.

b) The most likely area having been located, it is photographed using (say) a 1.5m Schmidt camera telescope in space. The plates are taken at intervals of a few hours and then scanned for signs of a very hot blue star growing brighter.

c) Phases a and b are backed by X-ray and uv data also taken from instruments operating in space. (The extremely hot dense pre-supernova star can be expected to radiate copiously at these short wavelengths.)

d) These search phases being successful and the supernova located before explosion, all instruments are brought to bear to study the rise in gamma- and X-ray, uv, optical, neutrino and radio fluxes as the star builds towards explosion point.

e) At the explosion point most (if not all) of the instruments are readjusted for sensitivity as the energy output leaps by about 1 billion times! At this stage gravitational and neutrino instruments are able to detect the oscillation of the central core and possibly predict whether a neutron star (Type I) or black hole (Type II) will result.

f) Finally, the instruments follow the stages of decay and can trace the evolution of a Type I or Type II supernova — depending upon which it is. Probability dictates that it is most likely to be a Type I event — in which case there is an immediate yearning for another supernova, this time of Type II, to occur!

Perhaps I am being optimistic in hoping that a Schmidt with a 1.5m mirror will be in space by 1999, but there is no real reason why not (apart from money and the will to do it). The following of a supernova from pre-outburst to as far as is deemed worthwhile, as I have described above, would yield data sufficient for decades of analysis and reanalysis. Knowledge of high-energy events on a scale we could never hope to copy would be obtained over a period of about five years, but our *total* knowledge and understanding would be increased on a scale which is difficult to quantify accurately. To say it would be doubled is probably a moderate underestimate.

To invest in placing such instruments in space and coordinating them is a small step, but to use them to study a supernova comprehensively would indeed be a giant leap for mankind — possibly more significant than the voyage of Apollo 11.

But, as shown in the next chapter, our prospects for doing just that are statistically being improved yearly by Nature herself. It is up to Man to keep pace.

The 'Ghost Supernova' in Cassiopeia

No chapter on supernovae would be complete without a mention of this object, discovered only recently despite its apparent age of 300 years or so. The first vague indications of a radio peak in the constellation of Cassiopeia were obtained by Grote Reber in 1940 during his sky surveys at a radio wavelength of 1.85m. The principal discovery came in 1947, during a sky search carried out at metre wavelengths, when a powerful radio source in Cassiopeia was identified. Since it was the first radio source located in that area, it was catalogued as Cas A, but the coordinates were not at the time resolved with sufficient accuracy to attempt correlation with an optical counterpart.

An accurate positional measurement was obtained by Ryle and Smith at Cambridge in 1950, and photography of the area by Dewhirst, also of Cambridge, triggered Baade and Minkowski to use the 5m Palomar telescope for more detailed photographs. These showed the characteristic 'shell' of filaments appropriate to the expanding remnants of a supernova, and Minkowski deduced from the rate of expansion that the outburst had occurred about AD 1700; i.e., that the supernova was only about 250-300 years old. This was a considerable surprise to astronomers, for to date no records have been discovered which relate to such an event.

Minkowski deduced that this object was close, by supernova standards, and estimated a distance of approximately 3,000 parsecs. Much of Minowski's original work has been followed up by Professor Sidney Van der Bergh of the David Dunlap Observatory, Toronto. The Americans finally concluded that the most likely date of the outburst was around 1667, and this was estimated from refined measurements of the rates of filamentary expansion in the supernova shell. Van den Bergh's most recent work has confirmed this date by the time-honoured method of going back through old photographic plates of the area of interest. In this case Van den Bergh was measuring rates of shell expansion over a longer period than that explored by Minkowski, and by examining the *high-speed* components he estimated that the outburst occurred between 1653 and 1671.

148

Current astronomical opinion on distance now places it at about 2,500 parsecs, slightly less than Minkowski's original estimate. It is not yet fully known if it is a Type I or Type II supernova but, from the abundance estimates made for various heavy elements, and from the nature of the expanding shell, several astronomers consider that it might be a Type II event.

If we assume this, then the supernova should have had an absolute magnitude of −18. If we take the distance to be 2,500 parsecs then the 'perfect seeing' visual luminosity would be reduced by about 62,500 times; this is a reduction of 12 magnitudes, so that the 'perfect' visual magnitude would have been −6.0. This is brighter than Venus at its peak, and about as bright as the Moon when 6 days old. Current opinion also suggests that the supernova occurred in a part of the Galaxy such that obscuration by intervening dust reduced the light reaching us by a further factor of 100, corresponding to 5 magnitudes. So finally we would see a star of magnitude −1.0, and this for a period of about 14-20 days. Then the magnitude would drop to about +1 or +2 and remain so for about 2 months. After a further month it would drop below naked-eye visibility and vanish from sight. We are therefore looking for an object that flared up and lasted as a new star for only about 3½ months. This was a miserly damp squib compared with the splendid 'new' stars of 1572 and 1604. But there should have been one outstanding characteristic about *this* particular supernova—it should have been a deep orange red because of the absorption of blue light by the intervening dust.

A bright orange-red star, appearing as an extension to the well known 'W' shape of Cassiopeia and brighter than all the surrounding stars should surely have been noticed — and yet so far no records have been found. Perhaps the short-lived event (as seen from Earth) took place during the summer months when the northern sky is brighter than at other times of the year; this tends to cause the circumpolar constellations (of which Cassiopeia is one) to fade and be lost. Other possibilities are that the peak activity took place during a period of full Moon — the bane of amateur sky watchers!

It is possible that the object may have been mistaken for a

comet — but there are no suitable comets listed. The brightest comets listed for the seventeenth century are Halley's Comet (appearances 1607 and 1682), a comet with a 70° tail (1618) and a comet with a period estimated as in excess of 8,500 years (1680).

There is also another possibility, based on the fact that this short-lived event took place during one of the most unsettled periods in Man's history. The most likely dates suggested span the years 1653 to 1671.

England, for example, experienced a Civil War in 1642-1649, knew a period of Commonwealth Parliamentary rule under Cromwell 1653-1658, experienced the Restoration 1660, underwent a Great Plague in 1665 and saw London demolished by the Great Fire of 1666.

Elsewhere in Europe the Thirty Years' War had just been concluded by the Treaty of Westphalia. This involved the Netherlands, Spain and Germany busily engaged in reparation. Germany later became allied with Britain in the Grand Alliance against France, which lasted beyond the seventeenth century and the most likely time of the supernova's appearance.

Even China was not exempt from turmoil during this particular period, for the Ming Dynasty which had started in 1368 came to an end in 1644, being succeeded by the Manchu Dynasty.

In the midst of all this upheaval, perhaps nobody but casual sky-gazers were interested in a brief celestial phenomenon!

There was another factor which may also have influenced matters — and this was the continuing use of *astronomical* features in *astrological* forecasts. A bright and short-lived 'new' star which was deep orange-red would probably signify war, blood and several other evils. Its short life would also possibly have been very gloomily interpreted. In short, if a court astrologer or soothsayer wished to enjoy a long life it probably paid him to not even mention the matter.

And yet, perhaps in some book or archive of old documents there may be the diary of some astrologer or soothsayer or physician which gives the clue. And there are references to unusual comets in the years around 1653 to 1671, apart from the record of Halley's great comet. In his work Daniel Defoe refers to a large comet observed in 1665.

A brief reference giving the correct position, brightness and duration would be satisfying confirmation of the work of Minkowski, Van den Bergh, Gull and others on this subject.

Is Cassiopeia a Black Hole?

At the time of going to press, the leading Russian astrophysicist Ioseph Shklovskii has offered a new explanation for there being no record of the Cassiopeian supernova which he suggests occurred in 1668 (i.e., two years after the Great Fire of London).

Writing in 'Nature' (p.703, Volume 279, 21 June 1979) he considers that the star was so massive that it immediately collapsed to form a black hole. This is consistent with modern theory, and also with ideas that the event was possibly a Type II supernova. Shklovskii proposes that the star either blew itself to pieces or collapsed out of existence. In either case, the normal violent explosion of a supernova is missed out, thus accounting for lack of observations.

As I mentioned earlier, the 'new' star could hardly have been missed, as it would have formed a bright orange red additional "leg" to the W of Cassiopeia, a well-known constellation.

Of the two theories offered, Shklovskii favours the formation of a black hole, and cites two main features of evidence. First, the enormous mass of the original star, which is estimated as 10 solar masses at the time of the outburst, and at least 20 solar masses when on the Main Sequence. No neutron star or pulsar was formed by the explosion, but the outer ring or shell of gas is estimated at 3 to 6 solar masses.

It is the composition of this expanding shell which forms Shklovskii's major evidence, for it is suspiciously deficient in iron and iron group elements. He suggests that at least 1 solar mass of these materials would have formed prior to explosion — and yet there is no trace of these materials in the filamentary remnants. He therefore concludes that a black hole of some 3 to 6 solar masses resides at or near the site of the Cas A outburst.

10 Stellar Remains That Surprise

In the previous chapter I outlined the remarkable characteristics of a supernova outburst and explained that a massive star is a super pressure cooker that converts hydrogen into far more complex nuclei. This process produces energy and thus enables a star to shine.

The energy production stops when the innermost core of the star has been converted to iron. This inner core collapses, as do the layers above it, but, as the whole mass of the outer layer commences to crash down, those immediately over the iron core attempt to explode outward, for they still contain energy-producing material. The explosion thus has two fronts or shockwaves, one travelling inward and the other moving outward. The outward moving shock front and the inward collapsing outer layers trap and squeeze the 'in between' layers. These are forced to take up less room than is normal and, as a result, the 'pinched' material moves rapidly through the iron phase and up into the still heavier elements. In doing so, it absorbs energy and ameliorates some of the crushing force and high temperature.

Detailed statements on *exactly* what happens are difficult, but these compressed layers are 'hustled' through the iron stage and on into nickel, copper, zinc, selenium, bromine, arsenic, molybdenum, silver, cadmium, tin, gold, tungsten, mercury, platinum and a host of other elements. This compression synthesis stops only when super-complex and therefore unstable elements arise, *and of themselves* not only damp down the inrush but actively oppose it. The result is that the outward travelling shock front becomes reinforced in a medium which is now falling off in pressure, temperature and density. The shockwave is thus able to travel faster and develop the kinetic energy to burst forth as the observed supernova explosion.

I personally think that this shockwave effect is more pronounced in massive stars, where the overlayers confine the outburst and allow such processes to work for longer and accumulate more energy. This might help to account for the 'shoulder' seen in the light curves of a Type II supernova, and also for the possibility of 'runaway cores'. These are rapidly travelling stellar remnants believed to be the result of a supernova explosion that was not absolutely symmetrical. This could easily be thought of in terms of the pinched layers exploding through a weakness in the outside layers. Once perforated, the rest of the material would start to escape into space through the 'hole'. The result would be like a colossal rocket effect which pushed the remainder of the star in the opposite direction. Since the escaping exhaust material would be travelling fast, the stellar remnant would also move rapidly. Several stars have been observed hurtling through our Galaxy at speeds approaching one and a half million kilometres per hour — and the energy source has been attributed to a supernova explosion.

In *Nature* (7 December 1978) Sir Martin Ryle and three of his colleagues, in a paper detailing the properties of what are believed to be supernovae, point out that one of the remnants (2013 + 370) might be moving at a speed of 5,000km per second. This is about 1.6% of the speed of light. The remnant concerned lies in the constellation of Cygnus — a veritable 'mine' of supernovae and other stellar occurrences.

From the above discussion one might argue that a Type II supernova would perform better as a synthesizer of the heavier elements, above (say) palladium (atomic number 46, atomic weight 106), and, since there are fewer Type II events, the elements above this weight should be less common. Examination of tables of element abundances in the Universe shows this, although there are a few exceptions to the general trend, to be the case. It would be naive, however, to cite this as conclusive evidence. What is really needed is detailed spectroscopic examination of nearby supernovae — and at present we are lacking such objects.

Betting on the Supernova Race

However, Nature is working in our favour. No supernova has been observed since 1604, and one that occurred in Cassiopeia about 1670 — Cas A — was not reported or recorded anywhere. If the Cas A event is included, an RMS figure of 257 years separates the dates of the outbursts. This is larger than the figure given on page 136, 150 years. This is because the latter figure is derived from a very large sample (over two hundred) of pulsars of varying ages, whereas the longer period is derived from a small sample of 9 comparatively recent events which (with one exception) were recorded; which were, anyway, visible from Earth. If we take the span of the dates involved (AD185 to AD1700) the interval of 1515 years would imply, from the 150-year figure, that there was at least *one* more event that was not recorded.

We might now ask: 'What are the chances of seeing a supernova by the end of this century?' Taking a mean period between events of 150 years, the least chance of seeing a supernova is about 1 in 8 while the best is 1 in 3. These are the two extremes of a 95% probability distribution, with median odds being about 1 in 5 to 1 in 6. I have deliberately refrained from being optimistic. Some authorities quote figures of 40 years as the mean separation interval between supernovae. On this basis, the statistical odds of an event occurring before the end of the century then rapidly approach 1-1 as average and about 2½-1 on as the most optimistic.

I do not have the confidence to quote *those* odds with certainty, but I *do* think that a good case exists *now* for considering outline plans for coordinating instruments which could observe such an event: some of these items already exist, others may be built before the outburst occurs, some may need to be built specifically *for* the event. However, I may be overdramatizing, and it is likely that if the event occurs before 2000 we would anyway be well prepared. There was every excuse for the ancient observers to be surprised by a 'new' star — there is none whatever for contemporary observers to be caught napping!

This view is strongly held by Professor Sidney Van den Bergh.

He has already circulated astronomers throughout the world with plans for observational requirements should a supernova occur within our Galaxy in an area where it can be easily observed. Ideally (and if the radiation levels allow it) I would suggest that the experiments should be reinforced by manned observations from suitable craft such as the Salyut and Spacelab series. Time permitting, the Space Telescope should also be available.

Professor Geoffrey Burbidge is often quoted as saying that 'modern astronomy is equally divided between study of the Crab Nebula and study of everything else'. If in the next few decades a supernova appeared in our Galaxy, Burbidge's quote might be rephrased as '99% devoted to the supernova, and 1% to everything else — including the Crab'! This would certainly be the case during the first 50 days or so. After that the observations would be less frequent until after (say) 5 years a small group would study the stellar remnant as a full-time project.

What would we gain by such a study?

Almost certainly a deeper understanding of the physics of nuclear processes, for we would have a multipurpose, high-energy, high-gravity synchrotron accelerator operating on a scale far beyond *our* capability and costing literally nothing to run! We may be given further insight into the optimized operating conditions for harnessing fusion power — again without having to build expensive pinch-effect (Tokamak) devices. Our knowledge of material behaviour under extreme relativistic conditions would be increased beyond the most elaborate dreams. If there *were* any hint as to forthcoming danger from radiation in times to come, then it may weld the various differing viewpoints into one common outlook — survival of the species!

Meanwhile we should follow up and develop Van den Bergh's plans and ideas for the event.

A Crushing Collapse

I mentioned that during the explosion shock fronts travelled both inward toward the core of the star and outward toward the surface. So far, we have considered only the latter, but what happens to the core of the star, which, as already pointed out, has collapsed to form iron? When the shockwave hits this area it cannot move out or expand as does the outer skin. The core just has to somehow absorb the energy, which we can assume to be half the total energy generated by the pinched layer explosion. Half the energy goes to the outside skin layers and half goes to the core. The latter is compressed to many thousands of times its normal density, but does not — and cannot — explode. Instead, it shrinks still further and has to give up this enormous energy by totally changing its form. Previously the iron core (even though superdense and superheated) *was* iron with 26 protons and about 30 neutrons forming a nucleus; furthermore, each proton would not trespass into the territories of its 25 other neighbours, and neither would the individual iron nuclei trespass on the boundaries of *their* neighbours. So, although of enormous density, there was still a comparatively large amount of space left in the core.

When the explosion wavefront arrives, the protons are *forced* to give up their privilege of extra space and the only way they can do this is to become neutrons themselves! This is done by the protons losing their positive charge (in the form of positrons) and giving up neutrinos. The positrons mix with the material inside the core and undergo several complex reactions releasing more neutrinos. These latter ghostlike particles are without mass or charge — and move at the speed of light. They remove the surplus energy from the core, thus allowing it to 'cool down' and stabilize. The core is now free to shrink, since all its constituents are neighbours. The degree of shrinkage is phenomenal. In the previous chapter I mentioned that a white dwarf with the same mass as the Sun had a diameter of about 30,000km, but if made entirely of neutrons the diameter drops to about 16km. The collected mass of neutrons has been variously described as 'neutronium', 'neutron soup', and in some cases 'quark soup', the latter name indicating that the material has undergone further consolidation. Although the theorists have attacked the problem from various angles, they seem to agree that a neutron star (the name given to the collapsed core remnant) has the following properties:

a) A diameter of the order of 16 to 20km with an outer eggshell-thick surface that is nothing more than tattered remnants of iron and heavier elements.

b) Beneath this outer layer, a further layer a few metres thick whose behaviour is totally dominated by the magnetic and electric fields above it. This crust, although essentially tightly packed neutrons, is so affected by these intense fields that it may behave as a metal, 10^{17} times stiffer than steel.

c) Beneath these two layers, a third layer of truly remarkable stuff — solid 'neutronium' with a density in the upper sections of 10^5 that of water; in the lower layer it approaches 3×10^{14} that of water.

d) The layer mentioned in c floats like a raft on the main component of the star, which is a superfluid 'neutron sea'. Towards the centre of this sea of superfluid the density rises to about 10^{15} that of water.

e) There is possibly a central core of solid neutrons or even quarks which — if the star is massive enough — may be on the point of liquefying.

The major part of the mass of the star lies in layers c and d — the solid and liquid neutron components. If our theories of neutron-star composition and size are correct, then enormous values for the densities in these areas must follow. An ordinary cube of sugar is about one cubic centimetre in volume; if made from neutron-star material it would weigh about 1 billion tonnes. Professor Harry Shipman, University of Delaware, has compared this mass to the total automobile output of the USA in a decade!

These figures seem utterly stupendous, but the astrophysicist is confident of his maths and logic. The main reasons for his confidence lie in being able to indirectly measure the size of the 'star' and directly measure its spin. This is so fast (from 1 turn every four seconds up to 33 turns a second) that the stuff needs to be of that density in order to provide a strong enough gravitational field to stop the star bursting as a result of centrifugal force.

The electrical, thermal, magnetic, and gravitational properties of these stars are also unique.

'It's Back!'

Those laconic words ushered in a discovery that has produced one of the fastest-moving frontiers in the history of astronomy. They were uttered in November 1967 by Jocelyn Bell (now Dr S.J. Bell-Burnell) and referred to the reappearance of a very weak but peculiar signal that she had noticed in August. In the normal traditions of the awkwardness of inanimate items, the signal had duly obliged by vanishing from any further records for the next six weeks! The reappearance confirmed its reality, and after an improved recorder had been installed the signals were seen to have a clock-like regularity.

In fact, the pulses *almost* appeared to be artificial rather than natural — an exciting possibility. After thorough checks to establish that the signals were not due to interference from adjacent channel broadcasts or electrical machinery, Bell and her tutor Professor Anthony Hewish began a long study of them. This eventually showed that the 'transmitter' was fixed in space at a considerable distance from Earth. There was no trace of the Doppler shifts of planetary axial rotation and revolution about the parent star. The signals were natural.

The first pulsar, CP 1919, had been discovered. (The letters CP stand for Cambridge Pulsar, and the number 1919 stands for Right Ascension 19 hrs 19 mins.) More work uncovered three more pulsars, thus confirming the phenomena as real and natural. CP 1919 has now been renumbered PSR 1919+21, the latter figures giving the declination and thus establishing a unique point in the sky, while PSR simply means 'pulsar'.

The list of pulsars has grown embarrassingly large since the

22 A tool already in use in Project SETI is the giant radiotelescope at Arecibo. Because of the reduction in background radio noise perpetrated by both the Sun and, on Earth, Man himself, the value of such an installation would be increased manifold were it to be sited on the Moon. The artist's preconstruction here shows three such antennae constructed within craters on the lunar farside. (*Courtesy NASA.*) See page 117.

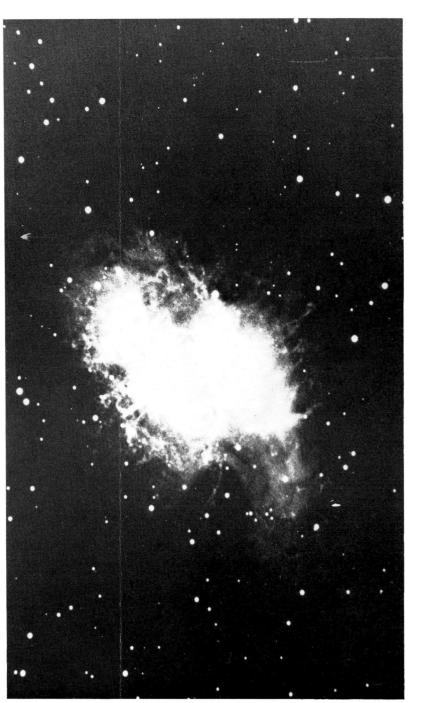

23 The Crab Nebula, at the heart of which lies the youngest known pulsar. The neutron star

first discovery in 1967; to date (January 1979) over 300 have been catalogued. They are situated at distances from 300 to 10,000 parsecs from us. They also appear to be of varying ages, the youngest so far observed being situated in the Crab and not quite 1,000 years old, while some appear to be several million years old.

Most appear to have only single 'humps' in the radio pulse, but others appear to have double or even triple 'humps' in their regular cycle. Each curve is the smoothed result of several tens or hundreds of individual records. Workers in the field become as familiar with these signatures as detectives with the faces, fingerprints and behaviour of their opposite numbers in the underworld.

No pulsars have so far been found with periods longer than 10 seconds and there are good theoretical reasons for believing that there is a cut-off point just prior to this critical period. The pulsar suddenly switches off its radio transmitter, although — as I mention later — it may still pulse in a different part of the electromagnetic spectrum. At the other end of the scale, the Crab pulsar 'beeps' merrily 33 times a second, and the third fastest pulsar, in Vela, performs 11.5 pulses per second.

As in human beings, the youngest are the fastest: the Crab has been identified with the supernova seen on Earth in 1054 while the Vela supernova has been estimated as having shone in Earth's skies only about 10,000 years ago: this means it was seen about 8000BC — and if, as we might expect, it was a brilliant object may have provided an astronomical stimulus to the early Sumerians, who were just on the point of developing the world's first written records!

Searches for pulsars in the supernovae of 1185, 1572 and 1604 have so far failed to reveal anything. This does not necessarily mean they are not there, but possibly that the radio beam does not sweep across the Earth's line of sight. If they *were* pulsars we would expect them to 'beep' about 50 times a second.

There is also a curious and as yet unexplained gap near 1 second where there are very few known pulsars.

The Crab Pulsar — and a Breakthrough

I have frequently mentioned the pulsar in the Crab Nebula, for here was a celestial Rosetta Stone awaiting discovery. This came in October 1968 when Staelin and Reifenstein of the National Radio Astronomy Observatory at Green Bank, Virginia, found a pulsar in the midst of the Crab. Here at last was firm evidence that pulsars were somehow associated with supernovae. This had been suspected as a result of the discovery of PSR 0833 – 45 (the Vela) by Large at Molonglo, New South Wales: it was known that the Vela nebula was a supernova remnant and PSR 0833 – 45 lay very close to the centre of it; discovery of the Crab pulsar clinched matters.

Now catalogued as PSR 5032 (originally NP 5032) this became, and has remained, a much studied object. It is comparatively near, it is well documented, it has a good strong signal.

What was it? This was exciting. The possibility of actually *seeing* a pulsar became real. This hope was based on some very tenuous theories of how the radio pulses were produced, and also on the emission generally observed at other wavelengths. Although no optical pulsation had been detected at that time, the object was known to be a strong emitter in uv and X-ray regions.

Also, a deep blue star with a faint but peculiar spectrum had been noticed by Baade. It was right at the centre of the nebula and Baade suspected it was the remnant of the explosion. Two teams (both in the USA) finally confirmed that this peculiar star *was* producing deep blue optical flashes which were in exact time with the radio pulses.

One of the experiments was extremely elegant and involved placing a television vidicon camera at the focus of a large telescope (the Kitt Peak 2.1m). With it was placed a spinning disc with a series of holes drilled round its circumference. The light from the star passed through these holes before being picked up by the camera tube. As a result the starlight was strobed. By adjusting the wheel speed and observing the television screen, the 'villain' star was unmasked. It was flashing on and off at a rate that made it easily discernible — and the star responsible was the one originally suspected by Baade.

There is another faint star above this optical pulsar but, although suspected as another stellar remnant, no pulsations of any nature have been detected to date.

Here at last was the long sought-for evidence, pulsars and supernovae were related; the former was all that was left when the fireworks were over! But to determine what the remnant actually *was* required the full unravelling of the characteristics of the pulses.

Neutron Stars Revealed

The flashes of light now provided an accurate measure of the *diameter* of the emitted beam of light, which in turn would indicate the diameter of the object emitting the beam.

Think of a light on the rim of a record player turntable as an example. Have this beam (ideally a laser) move past a photoelectric cell as the turntable turns, and measure the recorded shape of the cell's output. First there will be a low output, since the cell is dark. Then, as the sweeping light beam moves past it, the cell output will reach a maximum, stay there for a period and then fall back to the dark value again. Time the width of the pulse from half its maximum as it increases until it is at half its maximum while decreasing again. Knowing the speed with which the turntable spins, you can calculate the *angular* width of the beam. If the time taken for 1 revolution of 360° — i.e., one complete cycle — is T and the time taken to travel the pulse width is t, then the angle covered by the beam is $(t/T) \times 360°$. (In practice, this width is so small that the angle is measured in seconds of arc.) Using this principle on pulsars, if the distance of the object is known, then, knowing the angular diameter, the real diameter can be calculated. When this was actually done, the answer was *less than 20km!* The mass of the star having been estimated by other means, the density was calculated. It was so staggeringly dense that there was no doubt that pulsars *were* neutron stars. No white dwarf can possibly have this small size and colossal density; only the 'neutron proletariat' permits such extraordinary behaviour!

The searchers, having successfully seen optical flashes from the fastest pulsar, turned their attentions to the then second fastest — the Vela. And here they ran into problems, for the light output was less. Although it was spinning at one third the rate of the Crab, its light production would be reduced by a vastly greater amount; this was suspected from the way the radio output changed with rotational speed: it varied as the inverse of the *fifth* power of the spin period. Thus the Vela, spinning at about one third the rate of the Crab, would have an optical output of $1/3^5$ — i.e., 0.0041 — of the Crab's. In astronomical parlance this corresponds to about 6 magnitudes less. The Crab pulsar was already an extremely faint object at +18.5 *mean*, with the flashes slightly brighter (+17). The Vela was therefore expected to have a visual magnitude of +23 to +24 at its brightest, and slightly brighter flashes from the pulsating areas.

A suspect star and likely candidate was identified by B. M. Lasker in 1976. Using photographic plates that were highly blue-sensitive in combination with the superb new 1.2m Schmidt camera telescope at Siding Springs, New South Wales, he identified it as having a magnitude of +23.5 and therefore of the right order of mean brilliance. In addition it had a continuous spectrum, which meant it was probably *not* a normal Main Sequence star. It was close to the position quoted for the Vela pulsar and also near the centre of the nebula that was the remnant left after the main explosion. Lasker called it 'M'. Although it was suspect in 1976, the final unmasking had to await the development of sophisticated electronic equipment.

In 1978 an Anglo-Australian team (Peterson, Murdin, Wallace, Manchester, Penny, Jordan, Hartley and King) announced that they had confirmed 'M' as being an optical pulsar with a period identical to that of the radio pulses. As an additional surprise, though, the team found that nearly half of the output was unpulsed and had a strong red component. The reason for this difference in emission colour has so far not been explained. Detailing their work in *Nature* (30 November 1978) the research team compared the steady output of the Vela with that of the Crab and found that the latter too had a red component of steady light output. The level is very much less and is only about 2% of the pulsed light output.

We now have two firm determinations of optical pulsar characteristics. The main properties are shown in the table below.

Characteristic	Crab	Vela
Rotation rate (revs per sec.)	33	11.4
Mean steady absolute magnitude (M_B)	5.0	15.3
Distance (parsecs)	2,000	500
Age (as seen by us today)	1,000 years	10,000 years

I have deliberately kept the data to a minimum in the table, for in doing so a very clear contrast is given. The Vela pulsar is ten times the age of the Crab (so far as we are concerned — their absolute ages are irrelevant) and its absolute magnitude is down by 10 (a factor of 10,000 in brightness). A short calculation from these actual measurements indicates that the optical brightness changes as $1/R^7$ rather than $1/R^5$. Since the inverse $1/R^5$ law was deduced from fundamental physics and appears valid for radio work, there would seem to be additional factors at work (intermediate absorbing dust or gas) which further decrease the light output.

There is a general resemblance in the light curves of the two pulsars, inasmuch as there are two pulses per cycle, and also a resemblance to some of the *radio* pulsar curves. This supports the theory that the mechanisms for the production of radio and optical pulses could be, although probably not identical, very similar. The question is, what is this mechanism?

165

The Relativistic Dynamo

With the existence of neutron stars finally proved, it remained to identify exactly how the star converted rotational energy into light and radio waves. There are two major theories involved. One suggests that the beam is formed by the magnetosphere; the other that the radiation process itself is beamed. Theories of the first class are usually termed 'light-cylinder models', while those of the second class are termed 'surface-emission models'.

Neither model can explain the phenomenon in full detail, but the light-cylinder theory is probably the one most widely accepted. It assumes that the neutron star has a magnetic field whose axis is displaced from the rotational axis just as is the case with the Earth. The star spins amid a cloud of charged particles and drags the magnetic field round with it, thus causing it to slice *through* the particles. This produces large-scale currents amid the particles. (To appreciate the enormous forces at work, we should note that the magnetic field is about 10^{12} to 10^{15} gauss: the Earth's field is about 0.5 to 1 gauss, so already we are considering forces that are at least 2 million million times more intense, and possibly some thousand times greater than that.) The particles have been dragged towards the surface by the colossal gravitational pull of the star, but are eventually held off by mutual electrostatic repulsive charges. The voltages that exist above the surface of a neutron star are thought to be of the order of 10^{12} volts — about 2 million times higher than the voltages typically used on power grid lines.

We now have the conditions for generating electrical and electromagnetic power on an enormous scale, and this is precisely what a pulsar does. Low down in the particle atmosphere it is likely that the stirring of the soup produces large electric currents which (I would suggest) may do much to create the fine structure of the pulses by locally disturbing the magnetic field. It may also produce the steady light component noted in the Vela. Higher up in the particle atmosphere, the physics lies definitely in the relativistic area; electromagnetic effects are produced.

If the star spins 30 times a second and is 20km in diameter, then at the *surface* of the star the magnetic field is travelling at about 63km per second. As the distance above the surface increases so does the velocity with which the magnetic field slices through the soup of surrounding particles. At a critical distance the velocity approaches that of light. This critical distance can be shown to be about 45,000km if the pulsar is spinning at one revolution per second. Since the Crab pulsar is spinning at 33 revolutions per second, the critical distance is 33 times *less;* i.e., about 1450km.

At distances of about this the particles strongly resist the magnetic field and, in so doing, release relativistic energy in the form of photons. Depending on the strength of the magnetic field the photon spectrum may be radio, optical, X-ray or even gamma-ray. The drag effect so produced gradually 'brakes' the star and it slows down. This slowing down is observed with all pulsars, but shows to best effect when comparing the Crab and Vela objects. As the pulsar slows down, the diameter of the speed-of-light cylinder traced out by the magnetic field gradually increases. The strength of the field itself decreases as the square of the distance and these two effects combine to give an overall fall rate proportional to the inverse cube of the radius of the light cylinder. Finally, the slowing down effect itself obeys an inverse square law, and thus we have the overall $1/R^5$ relationship expressed earlier. It is because of this simple derivation from first principles (which works for radio pulsars) that results appeared so disappointing and initially fruitless when attempts were made to locate the Vela optical pulsar. Not until *ten years after the discovery of the Crab* were the efforts crowned with success.

The table below shows clearly how the output drops drastically as the rotation rate is reduced. If it were 10 parsecs away, we would have no difficulty in actually *seeing* the flashes from the Crab with the *naked eye* and a stroboscope; we would need a moderate telescope to see the Vela; while a pulsar with a pulse rate of two flashes per second would need a mirror the size of Palomar to see it even at this relatively small distance. Pulsars of periods 1 second and below would need a Palomar boosted with photo-electron amplifiers in order to obtain a record.

THE ROTATIONAL SPEED/VISUAL MAGNITUDE RELATIONSHIP FOR PULSARS AS SINGLE NEUTRON STARS

Pulse Period in secs	Rotation R in revs per sec	Fraction of 33 (f)	f^5	$1/f^5$	Diff. of $1/f^5$ (Mag)	Abs. Mag M_B	Vis. Mag. m_V at 2,000 parsecs
$30.3×10^{-3}$ sec	33	1	1	1	0	5	18.5 [a]
$50×10^{-3}$ sec	20	0.61	$8.2×10^{-2}$	12	2.7	7.7	21.2
$59×10^{-3}$ sec	16.9	0.51	$3.5×10^{-2}$	28.5	3.64	8.64	22.1 [b]
$87×10^{-3}$ sec	11.4	0.35	$4.9×10^{-3}$	203	5.77	10.77	24.3 [c]
$125×10^{-3}$ sec	8	0.24	$8.3×10^{-4}$	1200	7.7	12.7	26.2
$250×10^{-3}$ sec	4	0.12	$2.49×10^{-5}$	40,200	11.5	16.5	30.0
$500×10^{-3}$ sec	2	0.06	$8.1×10^{-7}$	$1.23×10^6$	15.2	20.2	33.7
1 sec	1	0.03	$2.43×10^{-8}$	$41×10^6$	19.0	24	37.5
7.6809 sec	0.13	$3.945×10^{-3}$	$9.58×10^{-13}$	$1×10^{12}$	30	35	48.0 [d]

[a] Crab Pulsar
[b] PSR 1913 +16
[c] Vela Pulsar (actual m_V 15.3)
[d] Norma X-ray Pulsar (actual m_V 19)

Most of the known radio pulsars are at an average distance of about 2,000 parsecs and the last column gives a general figure as to how bright an optical pulsar is as seen from Earth. It is possible that existing 'electronically boosted' ground telescopes will attain magnitude +30. Beyond that, the requirement can only be met by a space telescope: its ability to integrate over long periods and its freedom from background stray light permit 'staring' at an object for far longer than any Earth-based installation could.

Note that the table mentions 'single neutron stars'. Armed with such forecasting knowledge we could predict that it is unlikely that any other optical pulsars will be found. *That* forecast was proved wrong in December 1978! Another optical pulsar *was* identified. It did not conform to the $1/R^5$ rule. It was a different type of pulsar — but it, too, was a neutron star, and a true supernova remnant.

The Norma Optical Pulsar

The *New Scientist* of 4 January 1979 announced that observers of the European Southern Observatory at La Silla, Chile, had successfully detected light flashes from a star whose position corresponded to 4U 1626 –67, a pulsating X-ray source in the constellation of Norma (the Rule). The source generates X-rays with a periodicity of 7.6809 seconds for each pulse cycle. The optical flashes recorded by the La Silla observers correspond exactly in phase and duration to the X-ray pulses.

According to the 'light cylinder' theory, the optical output from a pulsar operating at this rate should be of absolute magnitude only +35 (visual magnitude at 2,000 parsecs, +48). This is way beyond the range of even the best present-day electronically aided telescope — and the 3.6m at La Silla *does* rank among the world's best. For the pulses to be recorded with the apparatus used (an EMI 2-stage imaging photomultiplier), the magnitude must be about +19 to +20 at most. The seeming disparity of about 30 magnitudes (or 10^{12} — 1 trillion) is explained by the X-ray pulsar having a *visible* counterpart. This is a companion star, and the mechanism involved is similar to that invoked for an ordinary nova; i.e., a stream of matter (hydrogen) is being syphoned off from one star and is falling down to the surface of the second. In this case, the accepting star is rotating fairly rapidly — 0.13 revolutions per second — and is almost certainly a neutron star. The donor star is visible and is therefore probably a luminous giant.

The flashes observed most likely correspond to brief glimpses of the neutron star 'hot spot' — the landing or impact point of the material donated from the companion. The process is more efficient than the light-cylinder — or synchrotron — reaction of the single neutron star, and may be considered as a quasicontinuous 'mini-nova'. We do not know precisely how long such stars stay in their pulsar condition, but recent work by de Loore and de Breve of Brussels University indicates it may be several million years.

The European Southern observers have obtained a 3-stage EMI photomultiplier, and it is highly likely that they will considerably extend their survey to cover other X-ray pulsars. One hopes they will amass statistics that will further supplement our meagre *optical* knowledge of neutron stars.

X-ray pulsars span a pulsed time bracket of less than 1 second to intervals of more than 1 minute. This area is of great interest for it may cover several modes of operation loosely grouped under the heading 'pulsar'. This includes single fast-rotating neutron stars, binary systems of which *one* is a neutron star and binary systems where *both* are neutron stars. In the last case, both stars may revolve inside a common atmosphere or magnetosphere and complete an orbital revolution in a matter of minutes.

Much of the deductions dealing with pulsars are largely theoretical and based on slender evidence. The supposed structure of an outer crust and a sea of neutrons is based on the observation of 'glitches', whereby the star slows down at a rate that is constant, then abruptly speeds up, and afterwards slows down again at its previous rate. The effect has been likened to a 'starquake' in the outer crust: this would undergo sudden deformation, relieving the stresses set up by the slowing down and allowing the crust to catch up with the faster-spinning interior. The model used is closely akin to that adopted for the Earth; i.e., solid crust/liquid interior/solid centre, and there is a known relation between earthquakes and minute variations in the Earth's rotation rate.

This model may be naïve but, with our presently limited knowledge, it is the best we can offer — further pulsar examination is required over the widest possible electromagnetic spectrum range.

Recent Discoveries

Two recent discoveries have tended to confuse the neutron-star picture rather than clarify it. *New Scientist* (12 October 1978) reported the results of a new survey carried out by the improved Mills Cross Radio Telescope at Molonglo, New South Wales. The team was an experienced international group — Andrew Lyne of Jodrell Bank; Joe Taylor, University of Massachusetts; Richard Manchester of CSIRO Australia; and John Durdin, Michael Large (finder of the Vela) and Alec Little of the University of Sydney. They surveyed over 60% of the sky from declination –85° to +20° during a three-month stint in 1977. 224

pulsars were detected, 155 of which had not been previously charted. (The total of known pulsars now stands at over 300.)

From this sample they estimated the total number of pulsars currently existing in the Galaxy — and arrived at a figure of 1 to 2 million. This is about four times greater than any previous figure and has implications of one pulsar being formed *every four years*. This is ten times as often as any previous 'most optimistic' prediction. From examination of supernovae formation in other galaxies the average accepted figure (for an average galaxy) is one supernova each 150 years: the optimistic estimates suggest one supernova each 40 years. These latest observations from the international team imply that the production rate is a whole order of magnitude more frequent. Two suggestions have to be considered:

a) Pulsars can be formed without the intervention of a supernova. Perhaps stars collapse quietly into neutron cores.
b) The exploding star may form more than one stellar remnant.

Also, of course, there is always the possibility that the results are wrong — but they will be cross-checked in subsequent examinations and surveys. At present, we do not know the periods of the 155 new discoveries; we do not know if there is another 'Crab' or 'Vela' fast pulsar awaiting confirmation by optical observations.

The second discovery is more confusing. In *Nature* for 7 December 1978 Sir Martin Ryle (Cambridge), J. L. Caswell (CSIRO, New South Wales), and G. Hine and J. Shakeshaft (Cambridge) discussed observations of a new class of radio star and listed nine candidates of which three had possible optical counterparts. *None of these items are pulsars*, either optical or radio, but two of them *are* associated with X-ray objects. In the article, Radio Source 1516 - 569 is associated with Cor X-1 and with a possible optical counterpart at RA 15h 16m 48.51s, Dec −56° 59′ 12.7″, while Radio Source 1909 + 048 is associated with W50X and with a possible optical counterpart at RA 19h 09m 21.32s, Dec 04° 53′ 53.8″.

171

Ryle *et al* are of the opinion that these two objects could well be the remnants of supernovae since the radio, optical and X-ray properties are consistent with that idea. They are cautious in this suggestion for, although *near* a supernova remnant in line of sight, one of the nine listed candidates has now been identified as a very distant extragalactic object — not a supernova fragment in the Milky Way.

However, throwing caution to the winds, there seems little doubt that they have discovered a new class of supernova remnant. A simplistic assumption is that a pulsar has not been detected because the radio beam of the pulsar does not sweep in our direction. Another theory suggests that pulsars are not observed until the hot plasma in the immediate vicinity of the neutron star has been dispersed thus allowing the radio beams to break through. We shall have to await the results of further observations and more research on these stellar remains. There are probably several further surprises in store.

But stellar remains with properties so unusual that 'commonsense' immediately calls them impossible appear to exist, and we shall discuss them in the next chapter. Before that, let us look at a unique star system that links pulsars and these 'impossible' entities, black holes. It is called PSR 1913 +16.

PSR 1913 +16

This is a unique radio object which has played a major part in substantiating Einstein's General Theory of Relativity. It is unique in being a radio pulsar in a binary system, and has been closely monitored since its discovery in 1974. It is a distant object and is within reach of only the largest telescopes: from the degree of signal dispersion obtained it is estimated that it lies approximately 6,000 parsecs from us.

Because it is a binary, the masses of the components have been measured accurately; the most probable solution of the equations of the observed motion is that the pulsar has a mass of 1.39 ± 0.15 solar masses while its companion is of 1.44 ± 0.15 solar masses. The companion could therefore be either a neutron star or a black hole. Both stars are close to Chandrasekhar's Limit; and the

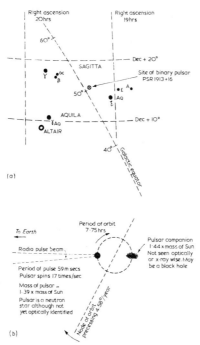

16 The truly remarkable binary pulsar PSR 1913 +16 (the Taylor-
Hulse pulsar). In the upper part of the diagram is shown the
position of the pulsar in the night sky; in the lower part is shown
the orbital configuration of the system.

authors of the paper dismiss the possibility of the companion
being a white dwarf. The system is shown in Fig. 16.

In this chapter we have been discussing the possibilities of
seeing the optical pulses from pulsars. PSR 1913+16 is one of the
fastest known pulsars, and should theoretically be visible to us,
despite its great distance. I have included it in the table on page
168, and, from the $1/R^5$ relationship, deduced an absolute
magnitude for it of nearly 8.7; i.e., it is six times brighter than the
Vela at absolute magnitude 10.77. The greater distance and the
greater brightness just about cancel out, so that PSR 1913+16
should in theory be almost as bright as the Vela pulsar which, as
we have seen, has now been optically located. The theoretical
visual magnitude difference is small — 24.3 for Vela, 24.6 for
PSR 1913+16 — and, since the amount of dust and obscuring

material between us and the binary is small, I would suggest mounting an immediate search for its optical counterpart.

If the companion also is visible, then it too is a neutron star. If it is not visible, it could be a black hole.

PSR 1913 +16 could be examined in detail over the next few years for it is a major clue in 'The Great Pulsar Mystery'.

11 New Cosmology Hides in Holes

Instruments above the atmosphere may assist in proving the existence of what seems to be the ultimate state of matter as we know it. Beyond this limit we cannot see, and possibly (there are reservations) beyond this limit we may not go. I refer of course to the black hole.

Black holes come in various sizes, and a very small one has recently been accused of drilling holes in Siberia in 1908. Since they have been judged possibly responsible for several other acts of celestial vandalism, it is not surprising that they are receiving considerable theoretical study, now beginning to be backed by observation. These objects may radiate strongly at wavelengths which are below the atmospheric cut-off, and hence orbital instrumentation will probably dominate in observing techniques.

Evolutionary Steps of No Return

Black holes are so freely exploited to solve awkward problems of cosmology that one may be forgiven for thinking that they are a recent invention or discovery. Not so. What *are* recent are the theoretical advances made in relativistic physics which enable us to gain a clearer insight as to the behaviour of matter under extreme gravitational conditions in combination with intense magnetic fields and rotational spin. These latter forces are almost — but not quite — as strong as the dominant gravitational field. Recently the brilliant work of Stephen Hawking has added the dimension of temperature to explain black-hole behaviour. My personal view is that we will probably have to add colour and charm to the relativistic equations that describe cosmic matter under extreme high-energy conditions. As observational instruments become more subtle and discerning, one hopes we may have to add further 'dimensions' to describe the forces at work.

The term 'relativity' automatically invokes the name of its

progenitor — Einstein. This is not entirely fair, for the tool that really broke the shell of black-hole theory was quantum mechanics, of which Schrödinger, Heisenberg and others who followed Planck's footsteps were the real founders.

However, in 1916 Einstein proposed his General Theory of Relativity, which explored in detail the behaviour of a body in a gravitational field, by considering it as being in a bent or warped curvature of 'four-dimensional spacetime', and the behaviour of the warp (the severity of curvature) by means of 'field equations'. Obviously the effect will be greatest in powerful gravitational fields, and in 1916 Einstein proposed tests that involved examining light rays from stars which passed close to the Sun; he suggested that these rays would be bent by the curvature of the spacetime warp of the Sun's gravitational field, and predicted the degree of curvature.

In 1919 a total solar eclipse provided the opportunity to do the test. Although the results were not perfect they were largely in agreement with theory. (Incidentally, although Einstein's General Theory of Relativity provides a more thorough explanation of gravitational field behaviour than do Newton's Laws of Gravitation, this experiment proved Newton partly right in his thinking that light had a corpuscular or particle nature. Up until Planck's suggestions of 'quanta' or 'wave packets', Maxwell's Equations of Wave Motion had dominated electromagnetic theory.)

Einstein, of course, had forged ahead and considered the implications of even more powerful gravitational fields, but did not publish any limiting solutions to the equations. This was first done by Karl Schwarzschild, who produced two clear and concise papers which showed that, if a body were sufficiently reduced in size and increased in mass, light would cease to escape from that body.

Wave a magic wand and reduce the Sun's diameter while preserving its mass. The gravitational field increases as $1/d^2$, where d is the diameter. Eventually the field becomes so strong that the Einstein spacetime warp bends right round and enfolds the 'Sun'. From that moment on it passes from view (except for a frozen image) and is lost from our Universe. But — it still has all the properties of a body with the mass of the Sun.

And what is the diameter at which all this happens?

From a magnificent golden ball 1.5 million kilometres in diameter the Sun has shrunk to being just below ten kilometres across!

I would here make a small and important point. Suppose that such a thing *could* and *did* happen, and that by some miracle we could survive it, what would we find happening to the dynamics and orbital conditions of the whole Solar System? *Nothing at all!* The Earth would still orbit at 1 AU from the black hole; likewise, all the rest of the planets would not 'notice' the difference gravitationally and would continue orbiting normally. We would, of course, be lost in Stygian darkness, so *we* would notice the difference.

It is, therefore, feasible that we could safely have a black hole of 1 solar mass near the Solar System without harm. We could, for example, quite happily endure the nearest major star being such an object. At a distance of 1.3 parsecs, Alpha Centauri is a beautiful triple star system: the major star is slightly larger than the Sun, the second is slightly smaller and fainter, the third is a dim red dwarf that needs a powerful telescope to see it. Any one of these could be replaced by a black-hole counterpart — and it would not affect us in the slightest.

Obese Neutron Stars

Karl Schwarzschild showed that a star which had a deadweight mass equal to three times that of our Sun could not avoid collapsing into a sphere about 15km in diameter (by 'deadweight' I mean zero energy production — no radiant pressure propping up the star's interior to stop this collapse). He also assumed that the object was not spinning and had no magnetic field — it was just an ideal sphere. It was an interesting mathematical solution, and there matters rested for about twenty years.

But perhaps some data from the previous chapter are beginning to stir in the memory? —15km in diameter? collapsed

177

interior? no energy production? Does this not sound like a neutron star? And, if that type of stellar remnant is found at the centre of a supernova explosion, is it possible that under certain circumstances we might find a black hole instead? Theoretically (if the iron/neutron core left after the fireworks is of greater than three solar masses) it *should* be a black hole, but there is an even chance that it will nevertheless be a neutron star. This is due to the effects of spin. We have already seen that an ordinary neutron star can spin at at least 33 times a second, so if the star were a little more massive we might be able to let it spin faster and allow the centrifugal force to prop up the surface against total gravitational collapse. With this faster rate of spin the 'neutronium' (even though it is 10^{17} times stiffer than steel) can behave very much like a ball of soft clay on a potter's wheel: it tends to fly outwards at the rim. Our proto-Schwarzschild sphere will tend to become a Schwarzschild doughnut (although without a hole in the middle).

If more mass is added to this elliptical doughnut, it can be made to spin faster and *still* stay stable. Eventually, however, if the mass is too great, a limit is reached where our neutron star either bursts at its equator and becomes more than one stellar remnant, or collapses at the poles of the axis of spin and becomes a black hole. Stars which may be in this condition are called 'obese neutron stars', and may go up to a limit of about 6.5 solar masses — i.e., twice the static Schwarzschild limit — and still survive.

To this I would add a note of my own in that, if such a star were a pulsar, then it might eventually undergo catastrophic changes as it was slowed down by the normal magnetic braking. Once below a certain critical speed the centrifugal force could not hold out against the gravitational infall. The mass of a pulsar remains relatively unchanged during its spin down so, at some time, one might expect an obese neutron star to either explode into fragments or collapse and, after shrinking through the Schwarzschild limit, become a black hole.

Black Holes have Shape but no Hair

In the above example of obese neutron stars, we have introduced a concept that was not considered by Schwarzschild — the property of spin — and, by using a simple analogy, have shown that extra force acting outwards would cause the neutron star to bulge at its equator, making 'room' enough for twice the mass of material that would be involved were the object stationary.

Now let us suppose that the obese neutron star *has* collapsed to a black hole. Will this be the doleful-looking round black dot so often portrayed on the pages of some textbooks? The answer is a definite 'No'. The star was spinning before the major part of it collapsed to form the hole, so the hole too will spin.

Since the hole will have a smaller diameter than the original star it will spin faster (since it has retained a large proportion of its original angular momentum). If the original neutron star had a diameter of 10km and rotated at 50 times a second, the black hole will spin at a rate that can be easily calculated. The angular momentum of a body is proportional to mr^2f^2 *where m* is its mass, *r* is its radius and *f* is its rate of spin. So, if our fat neutron star, of 6.5 solar masses and with a radius of 10km, collapsed into a hole having a radius of 0.5km, we can calculate how fast it will spin. If we assume the total mass is unchanged (i.e., *all* the material collapses into the hole), then the maths is simple: we multiply the original rate of spin by the ratio of the radii of the body before and after collapse. The radius ratio is $10/0.5 = 20$, and this multiplied by 50 gives 1,000 revolutions per second as the rotational rate of the black hole! An even more massive star could spin even faster, since the diameter of a black hole is increased only a little by even quite a large increase in mass. A black hole of 30 solar masses created thus could be spinning at a speed approaching 5,000 revolutions per second.

If a black hole kept its magnetic field intact during the collapse, and *if* its axis of spin does not coincide with the axis of the magnetic field, then *theoretically* at first glance it might seem to be able to produce pulses of light and radio energy in the same way as an ordinary pulsar. This would be exciting if it were real, for pulsars spinning at such frequences would, when picked up by our radiotelescopes, produce musical notes — or, to be more accurate, celestial shrieks: there would be little musical content in such emissions.

In this rush of imagination we are forgetting that supposedly nothing escapes from a black hole. Matter, light, radio waves and magnetic fields, if they still exist, are all trapped inside the spacetime warp. Indeed, this very lack of pulsar emission in what is obviously a massive collapsed star could be taken as supporting evidence for it *being* a black hole. So we can add another attribute: black holes are silent — they don't talk. There are no black-hole pulsars.

In my earlier discussions on spinning holes, I mentioned that, prior to its collapse, the star was probably an ellipsoid. As the black hole spins faster, outside the event horizon it assumes a different shape — it approaches that of a gramophone record: it becomes a disk. In fact, not all of the material may collapse into the hole: it may be left speeding around the rim and flatten out to form what is called an accretion disk. So part of the black hole is not round, but flattened, and it is surrounded by a whirling flat disk of 'ordinary matter'. If the matter from a nearby companion star were to be torn off and forced to spill down onto this disk, the violent conditions arising would produce X-rays. And it was as X-ray emitting objects that the first black hole candidates were found.

'Black holes have no hair' is a well known quote from Professor John Wheeler, who was pointing out the peculiar properties of these objects. Stars come in various sizes and some glorious colours — but black holes do not shine in the conventional sense. We could have a billion-tonne black hole that was made from exhausted star material, old boots, feathers, iron, water — anything that was material and had mass. It would always display the same characteristic: the ability to absorb light and not re-emit at any wavelength. Since we have no means to really get down to measuring black holes *in situ*, we are forced to the 'no hair' theorem, and have to make do with the low-grade information available.

Black Holes are Hot

The conventional view of a black hole being an object that kept a firm gravitational grip on everything in the vicinity of its 'event horizon' (the area of warped spacetime from which light cannot escape) was brought under suspicion in 1970 when it was discovered that this area always increases when additional matter or radiation falls into the hole. This holds true if two small black holes encounter each other and merge to form a large one: the total event horizon area of the single hole is larger than the sum of the areas of the two smaller ones — a strange case of 1 + 1 = 3!

There is one other factor — encountered throughout all physics — which behaves in a similar manner: it is called 'entropy', and the Second Law of Thermodynamics states that 'entropy *always* increases with time'. There is no way around this rule (life is the only exception), which may be likened to the water-steam cycle in an electrical power station. The water is boiled by adding heat and the resulting high-pressure steam drives turbines and generators. It gives up its energy and changes from 'lively', hot high-pressure steam to exhausted, low-pressure, warm water vapour that lazily drifts into a condenser to be reduced to water ready for reboiling. The thermal engineer states that, as the steam loses heat energy, its 'entropy increases'. So entropy is a measure of disorder, and time is the dimension in which it works.

Now let us take a slightly different example and put a nuclear boiler into our generating station. The atoms of uranium or plutonium split up in a fission process and release energetic neutrons and gamma rays plus light and heat. All of these degrade (i.e., gain in entropy) to give useful high-temperature heat energy. This boils the water to steam which loses *its* heat and gains entropy — and so on.

Since we are now dealing with nuclear reactions we can consider the largest nuclear furnaces — stars. In stars hydrogen fuel is turned to helium ash, releasing an enormous amount of energy and gaining in entropy. If the star is small, it becomes a bankrupt white dwarf gradually cooling off and gaining in entropy (disorder) continuously with time. But even if the star goes supernova, it still gains entropy: the disorder of the explosion is obvious and, if a pulsar is left, then that too slowly loses spin energy. If the explosion leaves a black hole then an even lower overall energy state is achieved — and the disorder is evident. This is the real meaning behind the 'black holes have no hair' theorem. If the material thrown in *is* highly organized (old boots, scrapped automobiles, etc.) we won't know by the information coming *out* of the hole because this 'information' is at the lowest possible level.

This is the crux of the nature of entropy at the event horizon area of a black hole. As a black hole consumes material its event horizon grows in area but cools down in an exactly analogous mode to the soggy water vapour exhaust of a power station turbine.

The 'no hair' theorem was further extended in the early 'seventies by various workers, principal among whom were Brandon Carter, Jacob Bekenstein and Stephen Hawking. Bekenstein was the first to breach the 'nothing escapes' theory and quantized the area/entropy idea by stating that, if a black hole had a finite entropy proportional to its event horizon area, then the hole had to radiate 'something' away from itself and that that something was in its lowest grade (worst disorder — greatest entropy); i.e., *heat!* And yet *by definition* the event horizon, the magic boundary defined by Einstein and proved by Schwarzschild, could not allow this to happen.

Here lay a pretty paradox, and one that remained unsolved until 1974, when the powerful tool of quantum mechanics broke the paradox. Hawking, mentioned often in these pages, decided to study the behaviour of matter close to the event horizon — the only major difference being that the spacetime laws would be those of quantum mechanics. To his considerable surprise he found that particles could move *away* from the event horizon — the unthinkable could be thought. The spectrum of these particles was thermal; the hole emitted radiation just as if it were an ordinary hot body. Moreover, it emitted from its surface with a temperature proportional to its surface gravity and *inversely* proportional to its mass. This means that, the smaller a black hole, the hotter it will be, and the temperature differences are extreme. A black hole with the mass of the Sun radiates at a temperature only one ten millionth of a degree above absolute zero. It would literally be a 'cold spot' in the 3 K background if there were no material in its vicinity (if there *is* material present then intense X-rays are generated as the matter crashes through the event horizon and vanishes). At the other end of the scale, a black hole of mass only a billion tonnes would have a diameter comparable to that of a proton and radiate at a temperature of 120 billion Kelvins. This corresponds to energy levels of about 10 million electron volts and the radiation would be partly gamma-rays but mostly electron-positron pairs, neutrinos and (according to Hawking) gravitons — these are the presumed quantum particles which carry away gravitational energy.

Einstein foresaw that gravitational energy would be lost by massive body interaction. His forecast has been fully confirmed by a recent experiment. The apparatus was in the form of the binary pulsar PSR 1913+16, 6,000 parsecs away, which we've already discussed. This binary pulsar is thought to be a rapidly spinning (16.7 times a second) neutron star orbiting another neutron star or black hole every 7.75 hours. As reported in *Nature* and *New Scientist,* both dated 8 February 1979, this faintly beeping object, first discovered in 1974, has been under close scrutiny for over four years. Using the pulsar as an accurate clock, J. H. Taylor, L. A. Fowler and P. M. McCulloch of the University of Massachusetts have monitored the changing period due to gravitational effects; to do this they used the giant 300m Arecibo radiotelescope. By allowing for magnetic braking effects and isolating other accountable items, this team has found that the pulsar is exhibiting a steady drift in orbital period which indicates that the binary partners are losing energy and slowly spiralling inward. Gravitational radiation has been predicted as being responsible — the system is transmitting gravitons, possibly from the black-hole component. The slowing down and nodal spiralling is seen as proof of Einstein's gravitational concept; but the system will still be monitored.

The gravitational effect is estimated as several thousand times that of the Sun. The latter's effect is to shift or precess the orbital node of Mercury by 43 seconds of arc per century. In PSR 1913 +16 the same effect nodally shifts the orbit round by 4.2 degrees per *year* — an increase of about 35,000 times!

So the combined forecasts of Einstein's predictions and Hawking's maths have been fully endorsed.

In at the Black, Out at the White

The full explanation of Hawking's emission theory may be simply given in terms of quantum mechanics by assuming that the *whole of space* is filled with 'virtual' particles. They are called virtual because they can be observed only indirectly. Nevertheless, they can be measured and their existence has been confirmed. They exist in pairs of completely complementary opposites. The effect of the black hole's event horizon is to pull *one* of the pair into the hole; i.e., 'space is broken'. This leaves the other particle of the pair without a partner. In this condition it has two choices — either to fall into the hole after its luckless companion or to rebound and recoil to infinity — and as may be expected on average the choice is split 50-50.

This phenomenon of burrowing out *against* a tremendous force field occurs elsewhere in physics and is known as 'tunnelling'. An electronic oscillating device known as a 'tunnel diode' relies on similar effects whereby the current in the diode material moves the wrong way, against the prevailing electric field. It does this because the lattice of the diode crystal is of the right dimensions to allow an electron to slip through and knock its neighbour out of place. This in turn progresses the wrong way — and so on: since a large number of electrons are involved we observe a current.

The analogy with the black hole is very close for we are merely substituting a gravitational field for the electric one. The gravitational field may be visualized as being 'grainy' with odd minor holes and patches here and there. The particle travelling out on the rebound is able to find these holes and slip through exactly as the electron moved against the prevailing electric field.

184

These concepts of 'graininess' in *all* structures and fields, and of virtual particles in what would be normally termed 'space', are probably now the twin pillars of much of our modern theoretical work. As Einstein displaced Newton, and Planck displaced Maxwell, so quantum mechanics will probably evolve into the long-sought Unified Field Theory. Armed with dimensions additional to magnetic, gravitational and spin parameters, and with a few more special quarks (fundamental particles), we may yet have an all embracing basis for explaining all observed physical phenomena. Einstein predicted this but died before any real developments or breakthroughs occurred.

Hawking's work has been confirmed by various approaches. It has also stirred others to explore more speculative grounds which usually start with the question: 'What happens to a Black Hole?' The answer to this depends on the hole's size. Small ones have short lives and, following Hawking's Rule, explode with some considerable violence. Larger ones, of solar mass, last almost for eternity (10^{66} years has been estimated — present age of the Universe is about 15×10^9 years), and larger holes than this last even longer!

Ultimately, therefore, the Universe may end as a series of large black holes; as they coalesce they form still larger — and cooler — holes. Entropy is at work again and steadily growing. Some authorities find this unacceptable and have suggested that the material aggregated in the black holes is somehow collected together, much in the manner of underground spring water. The matter then bursts forth into our Universe much as a 'gusher' of fresh water. This can be mathematically described by reversing the appropriate signs of the items that quantify a black hole; the theoretical product is not surprisingly called a 'white hole'.

Matter and half of the virtual particles should burst out of the white hole, while the other half of the virtual particles have the choice of leaping out or of committing suicide and jumping in *against* the outflowing mainstream. In theory, the virtual particles going down a white hole should balance those coming out of a black one, and, if space is to exist and possess the properties we attribute to it, then this balance most probably exists. One could then develop the arguments of virtual particles and space symmetry to argue that, if black holes exist, then white ones also must exist.

We are also forced to the conclusion that if (as Hawking shows) particles are tunnelling outward against a black-hole gravitational field operating inwards — and we label that inward movement as *positive* — then the white-hole conditions automatically invoke a *negative* gravitational field! This could be supported in theory for, if particles of ordinary matter found themselves in such a field, they would undergo mutual repulsion and literally explode apart at the speed of light, in an exact counterpart to the situation at the event horizon of the black hole.

We often read of the luckless astronaut who suffers the torture of being torn apart when he enters the crushing confines of a black hole; however, if we assume that he somehow survives, then he runs the equally fatal risk of being blown up on his way out of a white one! And the 'blow ups' would be complete: a complex of particles such as a man would be expanded and disintegrated by a force of $-10^{16}g$ with just the same rapidity and viciousness as he would be compressed and crushed with a force of $+10^{16}g$!

If black holes exist then white ones should also exist. If matter is being poured out of them explosively from a singularity or from 'somewhere else' then we should be able to detect them. Because they appear to represent enormous outbursts of energy occurring at immense cosmological distances, quasars have aroused interest as possible white holes. Unfortunately, they are far too distant to be firmly identified as yet. But, as we have seen, we can build larger telescopes, site them in space, and develop better image-processing techniques to look further towards the edge of our own Universe, and the improved sensitivity and resolution might enable us actually to see the structure of a quasar in detail. This would assist us in determining what they *may* be. At present there seems to be a family resemblance between quasars, Seyfert galaxies, and BL Lacertae objects.

But one strong objection to the theory that quasars are pouring energy into our Universe from 'somewhere else' is that it tries to circumvent the Second Law of Thermodynamics. It is the equivalent of perpetual motion, and suggests that there is a 'magic mechanism' somewhere that is winding up the Universe and preventing it from running down. In other words, the entropy is being *reduced* with time. This is contrary to physics — only Life reduces entropy with time.

Candidates for Black Holes

Once the existence of a black hole had become a near certainty in the minds of the mathematicians and conceptualists it was a foregone conclusion that astronomers would start looking for them. They were none too sure what precisely to look for since, by definition, black holes cannot be 'seen' in the strictest sense. They would be noted not for what they *are* but for what they *do*. Theoreticians had predicted that they would be characterized by:

a) being very small in both physical and observable angular diameter;

b) exerting possibly visual influences on their surroundings; and

c) being certainly the seat of high-energy physics.

Armed with these scant 'rules for black holes' the searchers set off. Not surprisingly, they tripped over the first candidate which offered itself without initially realizing what it was.

The seek-and-find episode started with the first X-ray satellite, *Uhuru,* launched in 1970. In March and April 1971 it sensed a change in the X-rays emitted by a known source — Cygnus X-1 (the first X-ray source discovered in Cygnus). This source had first been detected in 1965 but, because of the poor resolution of the detectors at that time, it could not be identified with any known optical object. The breakthrough came when it was noticed that not only had Cygnus X-1 changed its X-ray emission but that it was also emitting *radio waves.* Prior to 1971 radio searches of the area had proved conclusively negative; therefore a natural radio source had been switched on at the same time as the change in X-ray radiation had occurred. The radioastronomers could pinpoint this new source with the accuracy required for optical identification. Survey of photographic plates of the area showed the most likely candidate to be a hot blue star catalogued as HDE 226868. A further survey of this star showed it to be a blue giant about thirty times as massive as the Sun and about 30,000 times brighter. Knowing the real brightness and measuring the apparent brightness allowed the distance to HDE 226868 to be measured: it turned out to be nearly 2,500 parsecs. This enabled the 'candidate' to pass a *very* stiff test, for a rival theory had suggested that HDE 226868 was a nearby blue dwarf as opposed to a distant blue giant; had this been the case then the star would have been only about 200 parsecs away and the additional factors noted could not be attributed to a black hole but only to a neutron star or possibly a white dwarf.

Predominant among these extra features was the fact that HDE 226868 'wobbled'; i.e., it oscillated about a mean position with a period of 5.6 days. Astronomers are familiar with this phenomenon; it means that the visible star is a binary with an invisible companion. By measuring the degree of wobble and knowing the mass of the visible component it is possible to calculate the mass of the invisible one. In the case of Cygnus X–1 initial calculations showed it to be between 5 and 15 times the mass of the Sun, provided that HDE *was* a blue giant and that it *was* 2,500 parsecs distant. Comparison of this star with those of others in its vicinity during 1973 showed that HDE 226868's spectrum was that of a blue giant whose light had been red-shifted on its way to us by a long journey through clouds of dust and gas. This interstellar redshift favoured the distance of 2,500 parsecs rather than 200. The greater distance meant a giant and not a dwarf star, and that the invisible companion was an object whose mass was above that for a neutron star or a white dwarf.

The mass of the invisible companion has now received ever greater constraints and has been quoted as 8.5 solar masses, which places it above the limit set by 'obese' neutron stars. This recent figure was the result of work done in 1976 by Helmut Abt,

188

Paul Hintzen and Saul Levy. Using the facilities of the Kitt Peak Observatory (a superb 4m mirror and an equal-quality 2.1m mirror) they compiled 86 spectra of the visible star. They were thus able to state that, if there were a *third* component in the system, it could not have a mass larger than 1.5 solar masses and was therefore a white dwarf or neutron star. This is not sufficient to reduce the *secondary* component below the critical threshold level between an obese neutron star and a black hole. The team therefore concludes that the secondary companion is indeed top candidate for a black hole.

The major features of Cyg X–1 are as follows:

Visible component: Type B blue supergiant
Mass of visible component: 30 × solar mass
Diameter of visible component: 30 million km (approx.
    20 × solar diam.)
Mass of invisible component: 8.5 x solar mass
Diameter of invisible component: 15km (very approx.)
Distance of secondary from visible component: 20 million km
Orbital period of secondary about visible star: 5.6 days
X-ray brilliance of secondary object: 15,000 × that of Sun

Several key factors emerge from this summary. For example, these two massive components almost touch each other, the secondary skipping around just outside the main component's atmosphere. This ensures that the massive gravitational field of the physically minute secondary scoops off the outer atmosphere layers of the blue giant. This matter then accelerates down the gravitational slope into the 'hole at the bottom of the pit'. This allows the protons of the blue giant's atmospheric hydrogen to build up to energies of 100 million electron volts *per proton*. As these accelerated particles crash through the event horizon at the bottom of the pit they release X-rays with energies of 1,000 to 100,000 electron volts. The actual diameter emitting these X-rays is a hot spot smaller than the few kilometres of the companion, probably only 1km. This has been deduced from the flickering effect noted in the X-ray emission, which shows as rapid fluctuations of a few milliseconds' duration: light has a finite velocity, and the duration of these micropulses gives the diameter of the X-ray emission site, just as the pulse width of a pulsar gives the diameter of the source. The temperature of the hot spot has been measured as 10^8K.

So Cygnus X-1 is the prime candidate for a black hole and will probably be accepted as such — as may be several other similar items. Among these latter are X Persei and Circinus X-1. Circinus X-1 *seems* to be similar to Cygnus X-1, but the components have not been determined sufficiently clearly to say whether the companion is a black hole or a large neutron star.

X Persei is possibly an even more remarkable object, for it seems to consist of a visible star in partnership with a neutron star. This is emitting X-rays caused by the outer atmosphere of the visible star (a B-type blue giant) colliding with the surface of the neutron star. So far this is fairly normal, for over 150 X-ray systems have now been detected of which about ten are binaries. What makes X Persei extraordinary is that both the visible *and* invisible bodies appear to be orbiting a third invisible body — and this third body appears to be a black hole — with a mass 40 times that of the Sun. This is a tentative conclusion reached by examination of the X-ray curves produced by the system. The visible and neutron-star elements are thought to produce a characteristic binary X-ray period, but this is eclipsed at regular intervals and partially obscured by the binary pair orbiting the black hole. The presence of a black hole is deduced from the orbiting partner to the binary pair being invisible, and the mass of the hole is estimated from the duration of the orbital eclipse extinction period. If it is proved to be of 40 solar masses then the object certainly *is* a black hole — there is no question of such a mass being a borderline case. Again all the estimates rest on the assumption that the visible component is a B-type blue giant or supergiant at a few thousand parsecs. If the visible component is much closer then the relative masses of all three components drop drastically. But, even if the visible component were a blue dwarf only one tenth of the distance away, the invisible component would still be of 10 solar masses: unless the results were very accurately determined we might *just* make a case for an obese neutron star, but the general figures still indicate a black hole. However, accurate determination of the distance will be one of the first steps taken by Peter Stanford of the Mullard Space Science Laboratory, University College, London, who is heading a team making a detailed study of X Persei.

Another strange system is that of Epsilon Aurigae, a binary star of which only one component — a yellow supergiant — is visible. It is orbited by a dark companion which causes the visible star to be eclipsed every 27 years. During the eclipse period this star darkens from magnitude 3.4 down to 4.5 — a brightness reduction of nearly three times. All theories agree that the eclipsing invisible companion is a massive cloud of dust and gas surrounding a 'something'.

There are at least two theories that do not involve a black hole to explain what that 'something' might be. One considers that the invisible companion is a proto-solar system which has not yet condensed to form planets: we do not see the central star, the proto-sun, through the cloud of gas and dust. Another theory suggests that it is a rapidly rotating dull red giant, the central regions of which are being propped up by the speed of rotation.

A third theory considers it to be a black hole — and the reason for part of the brightness reduction is that the immense gravitational field of the black hole is 'stealing the light' from the visible companion during the eclipse period. We know from careful spectral measurements that the visible star is a yellow giant with a diameter of about 150 million km and a mass about 7 times that of our Sun. From this we can deduce that the invisible companion has a *minimum* mass of 8 times that of the Sun and a more likely value of 10 solar masses. This is definitely in the black-hole class, but any further evidence for the hole hypothesis is very weak. There is no measurable X-radiation or radio emission. One or both of these items would be expected if a black hole were surrounded by a dust cloud. Infrared and ultraviolet examination of Epsilon Aurigae over the next decade, involving orbital instrumentation, could well clear up the mysteries surrounding this intriguing star.

191

A more accurate assessment of the system has recently been carried out by Iwan Williams and Michael Handbury of Queen Mary's College, London. Initial results were published in *New Scientist* for 16 September 1976, and the full results in *Astrophysics and Space Science* for October 1976. These researchers estimate that the visible component of Epsilon Aurigae has a mass of 8 or 10 times that of the Sun and — most important to the hypothesis — *it is still contracting onto the Main Sequence*. This is directly opposite to the trend of all other theories which contend that the companion is a yellow giant *moving off the Main Sequence*, destined to become a red supergiant and eventually a supernova.

This move away from the conventional trends has enabled Williams and Handbury to consider the invisible companion as a disk of particles, each particle several centimetres across. The disk is semitransparent at the edges and slowly contracting. The total mass of this disk is about 10 or 11 solar masses and its central component is believed to be contracting and consolidating to form a star. The rest of the disk can then collapse to a planetary system surrounding this new sun.

The wide separation of the two components — 3.5 billion km or 23 AU (roughly the distance between the Sun and Uranus — is such that there will be little tidal disruption and, once formed, the planetary system will not be disturbed. This model is deemed by Williams and Handbury to be the best fit to the observed eclipses in terms of the overall light curve. This is a constant minimum, 330 days long, preceded by a 192-day decline and succeeded by a 192-day rise back to normal. The total eclipse period therefore is 714 days and, as stated previously, this occurs every 27 years (9,855 days).

Although extremely encouraging, this model is valid only if the star system is not more than 550 parsecs distant. The argument for this limit is that, if any further away, the stars would be so massive that noticeable evolutionary changes would have occurred in the century that the system has been under close scrutiny. Refinement of distance checks are an essential preliminary to the very close and detailed observation which will be done in 1982 — the year that the next eclipse commences. If the visible component is indeed still contracting then its temperature will be steadily rising and, as such, its ultraviolet output will rise more steeply than its visible output. In this context ultraviolet observation would be particularly valuable: if it were a disk of dust the invisible companion would scatter ultraviolet light; the ultraviolet extinction curve would therefore be much steeper than, say, the infrared one. A close scrutiny over a full eclipse period in both ultraviolet and infrared would provide a critical test for the dusty-disk Epsilon Aurigae model of Williams and Handbury. Unfortunately, the Large Space Telescope proper will not be in orbit by then. However, there will be several separate and smaller instruments operational, each of which will have the capability of adequately examining the star, which is a naked-eye object with a minimum magnitude of 4.5.

If this hypothesis is correct then Epsilon Aurigae is a *young* system and, as Williams and Handbury state, 'in the not too distant future this second star will become visible'. There is a precedent for this inasmuch as a very similar object — but at a more advanced stage — has been found in Cygnus. It is called MWC 349, is believed to be 1,000 years old and is a *glowing* disk of gas and dust of about 30 solar masses and 10 solar diameters. It is therefore much more compact than Epsilon Aurigae, and the energy of condensation has contributed to heating the material to the point at which it glows and is visible. Presumably the same will happen to Epsilon Aurigae and, in about 200-300 years' time (my guess), we may expect the invisible component to become visible. A third star of this type — RU Lupi in the constellation of Lupus the Wolf — also appears to have a companion which will become visible in the next few hundred years.

Epsilon Aurigae as a candidate for a black hole may have to be abandoned — there are better explanations for its behaviour.

In summary, therefore, we have very few black-hole candidates with the exceptions of Cygnus X-1, X Persei and possibly Circinus X-1 (but both of the latter items require further study). Cygnus X-1 alone is the only candidate that so far satisfies the mass, orbital, energy release, and size/density requirements expected of black holes. At this stage a strict stellar examination board would reject all other candidates as 'insufficiently qualified'. Further evidence is needed.

Mini Black Holes

The prime reason for a dearth of black-hole candidates is clear. Earlier in this chapter I discussed the possibility of black holes radiating particles and attaining a finite temperature in doing so. For black holes of solar mass the temperature was only 10^{-7} K (one 10 millionth of a degree above absolute zero). The black-hole candidate Cygnus X-1 radiates *thermally* with a temperature of about 10^{-8} K — because being more massive its event horizon is that much larger. It is literally an ice box in the sky. If we could point a super-discriminating radiotelescope at it and somehow miraculously isolate the radiation from that of the bright component we would theoretically 'see' Cygnus X-1. It would be a very cold spot in the sky, 300 million times cooler than the microwave background radiation at 3 K. Although such a 'thought experiment' immediately highlights the difficulties, it also points a way. Earlier I said that mini black holes radiated at 120 million degrees absolute; this is extremely hot, and radiation would be in X- and gamma-rays.

All of this has been predicted by Hawking, but he went further and showed that these mini black holes could have been formed when the Universe came into existence, about 15 billion years ago (according to current estimates). The shaped charge effect of such a stupendous energy release in the first few millionths of a second after the Big Bang could well have crushed nuclei together and formed mini holes in the primeval material. (Mini holes simply will not form in the interior of a small supernova. The temperatures and pressures, although high, are not high enough!) So a proportion of this primordial matter may still exist in the form of highly compressed mini holes. But the radiation with temperature is such that many of these holes must now be in the final stages of disintegration — by explosive energy release. Hawking noted that the energy release would be considerable, for black holes are very good energy converters. About 20% of the total mass of a black hole could be released as energy. Compare that with an H-bomb, which has a conversion efficiency of about 0.1%!

We can scale this to Hawking's billion-tonne black holes, the size of which he calculates would be about that of a proton but the explosion of which would be the equivalent of exploding a 2 billion megaton warhead! Hawking further calculates that there are about one million of these mini holes per cubic light year. Initially this large number looks frightening. But our chances of encountering a mini black hole are extremely small: on average 1 cubic light year holds 0.25 of a solar system (assuming all nearby stars have planets), but the actual *volume* of a solar system is less than 10^{-10} cubic light years (i.e., one ten billionth). So 4 cubic light years will actually house 1 solar system and 4 million mini black holes. Again, this looks frightening until we realize that if the solar systems were packed together edge to edge we could squeeze 4×10^{10} (40 billion) of them into the 4 cubic light years. This *then* averages 1 mini black hole per 10,000 solar systems. I for one am happy to live with odds of 10,000–1 against there being a mini black hole in our vicinity.

What is more important in Hawking's calculation is that the number of mini holes that may be exploding *now* should create a background signal detectable in the radio spectrum. Although the energy released in this area is lower than that in (say) the short X-ray region, the sensitivity of radiotelescopes is such that they offer the better chance for detecting mini holes. It is therefore hoped that searches can be initiated which will enable such outbursts to be found.

Very preliminary work has been done but no conclusive evidence obtained. However, the search will continue, for the possible presence of mini black holes has strong theoretical support. Success in finding mini black holes would indirectly confirm the already strong case for Cygnus X-1 and the developing evidence for X Persei. If small ones exist, larger ones must also — and the largest of all may be at the centres of galaxies.

Quasars — 'White' Holes that are Black?

Although discovered some fifteen years ago, quasars are still confounding astronomers and astrophysicists alike: how do these distant objects manage to shine with luminosities that exceed that of a normal galaxy and yet (from the observed light fluctuations) apparently be no larger than a few light weeks across? The power densities appear to exceed those of a supernova outburst but, instead of lasting a few months, quasars appear to be capable of radiating at these levels for several million years. Many theories have been evolved — and discarded. Early measurements of redshift which indicated that many of them were over ten times further away than the furthest other objects then known were eventually confirmed, and the distances at which quasars can be observed have steadily increased with improvements in optical-telescope sensitivity and photographic techniques. The limits of our observed Universe have been pushed to about five billion parsecs' radius, but already objects have been discovered which tentatively indicate distances of *ten* billion parsecs.

Two major theories are currently receiving attention, and both identify gravity as the energy source. They are the 'giant black hole' and the 'giant spinar pulsar' theories.

In the first idea a large mass (about 10 million solar masses) is being accreted and crushed in the same manner as in a normal stellar-mass black hole. In the second a mass of 1 to 10 million solar masses is again involved, but in this case it is spinning relatively rapidly and is being propped up internally by rotational forces and immensely powerful magnetic fields. The object therefore generates light and radio waves much as does an ordinary pulsar. This idea has recently gained credibility in view of the very regular flickering of light output exhibited by some quasars.

Work done in the USSR in 1976-77 by L.M. Ozernoy and V.V. Usov seems to provide evidence for intense magnetic fields and rotation forces in four quasars (3C273, 3C345, 3C446 and 3C454.3). All of these show signs of regular variability. (The rather ugly term 'spinar' has been coined for an object with these characteristics and which relies on rotational force to stop it collapsing.) Ozernoy and Usov point out that pulsed regularity is usually a sign of rotation and in their paper, published in *Astronomy and Astrophysics*, Vol. 56, 1977, they contend that this regularity weighs heavily against giant-black-hole models as the basis of the enormous energy output of a quasar.

Earlier in this chapter I mentioned that 'black holes don't talk' but that they *could* be rotating very rapidly. I also mentioned that black holes most probably have a large flat accretion disk. Now there is nothing that says this disk must be stationary — in fact, quite the opposite: the disk *has* to rotate rapidly to save itself from being gulped at one mouthful by the ever voracious hole at the centre! There is a fine balance which allows just enough material to trickle into the hole and no more, for the gravitational field of the hole itself is rotating and, as well as trying to engulf the material in the disk, it also grips it so firmly that the disk spins almost as rapidly as the hole. It is also self-stabilizing for, if the hole gulps down too much material, its gravitational field increases and grips the disk even harder. Not only this, but the matter falling into the hole makes *it* spin faster and with the increased grip the disk also spins faster — and creeps away from the hole.

If we take the analogy of a potter's wheel, let us imagine we are trying to paint a broad band of glaze onto the rim of a fired plate spinning on the wheel. If we pour on too much paint, it splashes and flies all over the place and a regular stable pattern is lost. If we apply too little we have a broken and irregular smudge. By applying the paint in the right amounts we match the paint flow to the wheel speed and achieve the desired result. A black hole with an accretion disk does just this for itself. Provided there is no external upsetting agent it will take just the right amount of material to stabilize itself.

Now let us consider a giant hole at the centre of a quasar. It has an accretion disk — but one with a difference: the disk may have sufficient mass and rotational energy to support a magnetic field. If the disk material is dragged through this field by the rotating hole, the system will broadcast light and radio waves. In other words, a rotating black hole at the centre of a quasar could be responsible for the observed variability.

If the fields and rotational speeds can be accurately assessed then it should be possible to forecast if the radiation is of optical, radio, X-ray or gamma-ray frequency. It would also be interesting to know if quasars which appear to have regular variability in output show any tendency to follow the $1/R^5$ rule deduced from the 'light cylinder' models of ordinary pulsars.

The 'black-hole motor' theory as regards the power source for quasars appears to be the contemporary favourite, and an excellent description of it was recently outlined by Martin Rees in the Royal Astronomical Society's Halley Lecture for 1978. Rees commented that, with massive objects (10^8 solar masses) like quasars, black-hole theory implied that the initial collapse took place so swiftly that little or no nuclear energy was released in conventional stellar reactions. If, however, the hole had an accretion disk, matter dropping into the hole could be converted into radiation or fast particles with a much higher efficiency than in any other nuclear process.

Assuming an efficiency of 10–20% (i.e., 100-200 times better than the hydrogen-to-helium process), then a quasar with a luminosity of 10^{46} ergs per second (a typical figure) would require only one or two solar masses of 'fuel' per year to keep it sustained at that rate. If the central mass and disk were of the order of 10^8 solar masses then the process would last for several millions of years. Rees went on to show that such massive objects *do* appear to lie at the heart of several celestial objects, including quasars and galaxies. Our own galaxy appears to have a central region which is a very peculiar compact radio source. It is not a pulsar, it is not a supernova remnant. It is only 10^9 km (a few light seconds) across and might be a region where there is a low level of accretion onto a massive black hole. Rees comments: 'It is a unique source in a unique place.'

12 Future Astronomies — Vistas in Cosmology

So far we have discussed the immense strides made in the branches of radio, optical, infrared, ultraviolet, X-ray and gamma-ray astronomy. Although not conventional astronomy in the accepted sense — i.e., the ray-defining and collimating mechanisms used are not mirrors or lenses of conventional format — the equipment nevertheless performs the same function. All of these astronomies (without exception) are dealing with electromagnetic radiation.

But in the last two decades two other branches of astronomy have been conceived that have nothing to do with electromagnetic radiation — although their subjects may (and almost certainly do) form an overall part of a Unified Theory. These branches are neutrino and gravitational astronomy.

Although the measuring instruments presently devised have been called 'telescopes', they are more correctly referred to as 'detectors' for they cannot 'see' in a directional sense but can only detect when a neutrino or gravitational event has occurred. The instruments presently developed, although large, are less effective than Galileo's first crude optical telescope.

But we must not be too hard. These detectors are large because they are dealing with very weak interactions. Neutrinos do not react with matter as does a proton or electron but can theoretically pass through about three parsecs' thickness of lead — or a complete galaxy of hydrogen — without pausing for interaction. Gravitons (the particles or quanta believed responsible for gravitational attraction) are not produced in quantity until a large mass is concentrated into a very small radius (i.e., when a black hole or neutron star is formed). Gravity is about 10^{39} times weaker than electromagnetic forces — and, apart from everything else, one's observations are dominated by the relatively large local gravitational field of the Earth — so it is not surprising that observers have difficulties.

In spite of these problems, the effects of neutrinos, antineutrinos and gravitons *have* been detected and, with a little encouragement, these 'cuttings' should properly take root in astronomical soil and grow in their own right into very healthy branches. The major fields in which they can be expected to flourish will be those of observing supernova outbursts and black-hole generation. All the electromagnetic branches deal with the *outside* of the outburst — the heat, light, and X-rays, etc., that are generated — but neutrino and gravitational astronomy should tell us what goes on deep down *inside* a star, right at the core of the reactions.

We expect neutrinos to tell us about the increasing temperature and energy levels being generated within a star. As the complexity of the nuclei being formed increases so the flood of outgoing neutrinos steadily rises, until a tidal-wave outburst occurs just prior to the core collapse. The collapse itself creates a tremendous surge of gravitational energy as the core changes from a compressed ball of iron plasma at a temperature of 3 billion K to a solid/liquid ball of neutrons at an even higher temperature. The ball has changed its diameter from about 15,000km down to about 15km for a Type I event. As shown on page 212 in a Type II the change may be far greater.

But the gravitational field and pulse energy are *inversely* proportional to the diameter *squared!* So, when the diameter changed by a factor of 10^4, the gravitational field and energy release increased by a factor of 10^8. (The actual figures may possibly be higher, for this simple calculation assumes that the core mass is unchanged during collapse; possibly additional matter may be pulled into the final collapsed core). The pulse is so large that a sensitive gravitometer should hope to detect it.

Furthermore, we may find that the core will 'ring' like a bell: instead of a single pulse of neutrinos and gravitons there will be a damped train of oscillation, which will decay exponentially to zero. This behaviour can be treated mathematically. When dealing with such wave trains, the ringing is more violent and non-linear the greater the excitation (mathematically, 'the greater the forcing function'). One can show this by whacking a bell really hard and hearing it screech as it generates overtones.

So, what happens if a really big star explodes and a larger than normal core collapses to form a black hole? I can understand a small star near the black hole limit collapsing quietly with a sigh — however, I instinctively feel that a large core collapse will be violent and swift. But a black hole by definition cannot allow a damped oscillation to occur in the ordinary sense of material rebound — for nothing escapes the event horizon once so formed, unless the *event horizon itself changes and oscillates*. The changing temperatures thereof (*à la* Hawking) allow the excess energy to be dissipated as heat and stability to be achieved by increase in entropy. I feel there is room for both theoretical and practical work in this area, for the collapse of a Type II supernova involves just such a massive core. Measurement of the gravitational and neutrino fluxes and *notes of change* of such fluxes would be invaluable.

Theoretically, then, we should (in the future) be able to devise a system which gives a series of signals and a data input long before (approximately four days) the optical outburst takes place. First the neutrino signals rise exponentially until the collapse; during this time the gravitational and optical signals are below recordable level. Then, at the outburst, both neutrino and gravitational signals show an oscillating train of pulses which die away exponentially: this is the core collapse. About 3-5 days later the optical brightness builds up, and we see the supernova. This increased brilliance may give us 300-600 days of naked-eye visibility — depending, of course, on distance and the extent of obscuration by dust between us and the outburst.

Detecting Ghosts

A neutrino is a 'convenience particle', for it was conceived to explain the observed energy differences occurring in the electrons emitted by substances undergoing radioactive decay. In theory — up to that time — all materials giving off electrons gave them off with equal energies, even though the reaction might be radium deteriorating to lead or thorium to bismuth. There was nothing absurd in this initial conclusion for, if the mass of the electron is known and the voltage required for the electron to escape from the parent atom also known, the two multiplied together should give the energy imparted to the electrons in 'electron volts'.

Only, it didn't work. When these energy levels were finally measured with indisputable accuracy, the results not only differed from theory but they differed by different amounts. Finally, in 1931, Wolfgang Pauli proposed that, when radioactive decomposition involved electrons, *another* particle was released simultaneously with the electron. This second particle accounted for the differences between calculated release energy and observed release energy. The theory only worked if the particle had no mass, no charge, a minute cross-section, and moved at the speed of light. In order that differing amounts of energy might be absorbed and removed from the scene Pauli proposed that the particle had 'spin'; i.e., that it could rotate at various rates and remove energy as nucleic angular momentum. In addition, he proposed that this particle had an equal and opposite antiparticle, for he foresaw that *all* particles should have equal and opposite antiparticles. (This basic principle of parity is still regarded as valid, although there are special circumstances where it is violated.)

Pauli's particle accounted for the energy discrepancies and allowed considerable progress to be made, but several of his colleagues regarded the proposal as nothing more than a mathematical fiddle used to make observed results square up with theory! Enrico Fermi thought differently, and immediately christened the particle the 'neutrino' (little neutral one). But the cloud of suspicion was not removed until 1953 when Cowan and Reines, using special tanks filled with water and a small quantity of cadmium chloride, proved the existence of antineutrinos. The water 'scintillated' or emitted a tiny flash of light whenever it was struck by a subatomic particle, but the cadmium also played a major part in almost immediately absorbing any *neutrons* formed while still allowing the antineutrinos to interact and produce two gamma-ray photons. Placing these special tanks in front of a heavily shielded reactor whose radiant flux was 10^{18} *antineutrinos per second,* Cowan and Reines were able to measure 70 antineutrino interactions per day; i.e., an average of 1 per 20 minutes (roughly 1,000 seconds). This means that, using suitable targets, we may hope to trap or stop 1 antineutrino particle in 10^{21}. This number is so large it is akin to comparing the Sun to all the stars available in all of the telescopically observable galaxies! But, incredible though the results may seem, the experiment was repeated and produced consistent results. Thus, twenty-five years after their conception, the antineutrino and, hence, the neutrino were proved to really exist.

Neutrino production is not tied purely to fission processes, for electrons (and neutrinos) are given off in fusion reactions. The Sun obtains its energy from fusion of hydrogen to helium. For every atom of helium produced, four atoms of hydrogen are consumed and two neutrinos ejected. One can calculate that, with the Sun shining at its present rate, 1.8×10^{38} neutrinos are produced each second (note that these are neutrinos, *not* antineutrinos). They are not obstructed by the enormous mass of the overlying materials but simply scoot through the Sun's layers until they are in free space.

However, they *can* be caught, but by a different reaction from that used in the antineutrino experiment. In the antineutrino experiment described earlier a material rich in protons (water) was used. When an antineutrino (symbol $-\nu$) was absorbed the proton (p) became a neutron (n) and a separate positron ($e+$). The physicist writes this as: .

$$-\nu + p \;\leftarrow\; n + e^+.$$

So, when dealing with neutrinos (symbol ν), we need something of a reversed process; i.e.,

$$\nu + n \;\leftarrow\; p + e^-$$

We therefore need a material rich in *neutrons* which can be turned into protons and electrons. There are, theoretically, several materials which can be used. One is 'heavy water' (deuterium oxide), the hydrogen of which has an extra neutron to its atom. When bombarded by neutrinos this neutron is temporarily turned into a proton.

The problems are that not only do we require a minimum of

about 500,000 litres of heavy water for reliable results but, even if we could obtain this quantity of what is a very expensive liquid, we would *still* have erroneous results. This is because the Earth itself is a prolific producer of antineutrinos, chiefly through the breakdown of radioactive uranium, thorium and potassium, and the total contribution from these materials is about 2×10^{26} antineutrinos each second; these would react with the protons our reaction has produced. The total emitted solar neutrino flux is about 8×10^{28} as intercepted by Earth. Now, even although this is about 400 times larger than the Earth's antineutrino flux, it is not much use if the liquid scintillator used intercepts *both* types of particle — and this is what would happen if heavy water were used.

Thus a material both cheap *and* highly selective to neutrinos alone would be very desirable. An Italian physicist, Bruno Pontecorvo, pointed out the particular virtues of the halogen group of elements for use in neutrino detection. They lie in Group 7 of the Periodic Table. Of the halogens, the first two are gases, the third is liquid, the fourth a solid, and the last a radioactive unstable solid which has been likened to 'very strong iodine'.

Group 7 Halogens	Atomic No.	**Group 0** Noble Gases	Atomic No.
fluorine	9	neon	10
chlorine	17	argon	18
bromine	35	krypton	36
iodine	53	xenon	54
astatine	85	radon	86

All are extremely poisonous, although three of them — fluorine, chlorine, and iodine — are essential (in minute quantities) to life. All of them attack metals and directly form salts (hence the derivation from the Greek *hals* — 'sea' or 'salt'). On the other hand, the 'noble' gases, in neighbouring Group 0, are indeed gaseous — even radon which, if not a gas, would be heavier than most metals, including lead, gold and tungsten. They do not easily form chemical compounds — one of the very few exceptions being xenon hexafluoride — and can therefore easily be stored for measurement.

Pontecorvo's great contribution was to point out that the halogens were rich in neutrons (fluorine has 5 isotopes, chlorine has 10, bromine has 5, iodine has 5 and astatine has 4), and that if a neutrino entered these complex nuclei a 'beta decay' would occur. This would have a twofold effect: a flash of light would be produced; and the neutron would be converted to a proton, the atomic number raised by one, and an atom of the appropriate noble gas produced. But this was not all. The atoms of gas so produced would be radioactive, and hence could be identified and counted with a high degree of precision.

Pontecorvo worked his way through Group 7 and dismissed all except chlorine. The solids were discarded on the grounds of cost, storage and difficulty of collecting the gas trapped in the solid; fluorine was expensive and dangerous to handle in gaseous form and, when he was assessing the materials then available, there were no fluorine-rich liquids (nowadays trifluorethylene, trifluorchloroethylene and tetrafluoroethylene are manufactured in quantity — but are not cheap). Chlorine was available in a wide variety of liquids, the cheapest and best of which were carbon tetrachloride and perchlorethylene. Both of these substances have a similar composition: four chlorine atoms, bound together by a common carbon atom in the case of carbon tetrachloride and by two carbon atoms in the case of perchlorethylene. Both are commonly and cheaply available as commercial, noninflammable, noncorrosive cleaning fluids, and the quantity handling techniques are well established. And quantities *are* called for in neutrino astronomy — a minimum of 450,000 litres was the calculated requirement, with 2 million litres preferred for more certain results.

Thus was the 'neutrino telescope' (more correctly referred to as a detector array) conceived in principle by Bruno Pontecorvo. Initially, however, it was put into practice by Americans: Raymond R. Davis and his team are currently operating the largest installation in the western world, but prior to this, in 1965,

Reines, whom we have already mentioned, reported that he had detected 7 neutrinos in 9 months. These neutrinos were of extra-terrestrial origin but could not be firmly attributed to the Sun.

Now, after more than thirteen years' work at the Homestake Gold Mine at Lead, South Dakota, Davis and his collaborators from Brookhaven Laboratory still have not been able to identify a positive solar neutrino flux of the expected value. To eliminate the possibility of spurious reactions arising from energetic particles other than neutrinos, the detecting fluid — 450,000 litres of carbon tetrachloride — was placed 1,600 metres underground, down a disused mineshaft; the galleries and chambers fanning out from the shaft are crammed with tanks and detecting gear. Each tank contains a set of scintillators and photomultipliers to register the gamma-ray photon flash produced when a neutrino event occurs. Failure to detect the neutrino flux is just as valuable as a positive result — if not more so!

Had the neutrinos been detected as expected, we would have been well content with our theories and models of the solar interior. But the experiments carried out by Davis have been repeated elsewhere with the same result — a low neutrino count. The surprise of this outcome has been well matched by the flurry of papers attempting to explain what has happened. The ideas covered include some that are rather startling. At present the question is wide open. The ideas put forward include the following:

a) The Sun has a black hole at its centre. This accounts for a large percentage of the internal mass normally attributed to helium. The core is therefore cooler than expected and the neutrino reaction, being temperature-sensitive, is thereby reduced in intensity.

b) The Sun has a *larger* quantity of helium than suggested in earlier theories. The larger core is cooler than expected and neutrino emission is less.

c) The Sun has temporarily 'turned off'; i.e., stopped producing helium. It would probably be more accurate to say that it is running at a lower power level, below the detector threshold.

d) The neutrinos *are* being produced but they are below the energy level required to initiate the chlorine-argon reaction. This means that we could not detect any solar flux however big the tank was.

e) The bombardment of chlorine by neutrinos is producing not gaseous argon but solid waxlike polymer particles which sink to the bottom of the carbon tetrachloride. This could easily happen if the fluid were contaminated by minute amounts of water. Unfortunately, the amount of wax produced is so tiny (a few hundred molecules per year) that it could never be recovered.

Theories *a* and *b* promptly contradict each other, and *b* especially is not convincing to the school of thought that considers the steadily growing central core of helium to be growing hotter, not cooler. The two most rational explanations seem to be *c* and *d*. In turn, *e* is improbable because Davis and the Brookhaven Group tested the system with a fission reactor at Savannah River using 4,500 litres of cleaning fluid. Unfortunately a fission reactor produces antineutrinos and the test relied on the results being 'double negative'. The antineutrinos do not produce the chlorine-argon reaction; therefore, if there is no output count, the system is suited to neutrino detection! And such was the case: the antineutrino flux did not cause a reaction in the carbon tetrachloride and hence the system would detect neutrinos.

Until we get a fusion reactor working we will not be able to fully check a neutrino detector with particles identical with those produced by the Sun. It is not enough to produce any type of neutrino — only those produced by a hydrogen proton decaying to form a helium neutron are acceptable.

An alternative form of detector is therefore desired — and it is possible that Nature has provided at least one (and maybe two) neutrino detectors. Both are true neutrino detectors and use the chlorine-argon system, but the chlorine is locked up as salt (sodium chloride). One detector is liquid (the sea) and the other is solid (rock salt).

The latter may turn out to be doubly useful, for I would suggest that cores of rock salt taken from a sample a thousand metres or more below the surface of Earth could have been affected by neutrinos — but by little else. Since such subterranean salt was buried at some past epoch it is possible that a potted history of neutrino activity may be available. The date would be as accurate as that for the surrounding rock. The salt, being a hard crystalline solid, would also possibly trap the gaseous argon and hold it until the core were dissolved in ultrapure water. The neutrino events recorded by 450,000 litres of carbon tetrachloride are approximately ten per year, and, since the liquid occupies a volume of about 280 cubic metres, this gives an average of one neutrino event per 28 cubic metres per year. If a core of rock salt is drilled out from the Earth then, depending on the recovered length and core diameter, about 2 to 20 cubic metres of salt could be recovered from each core. Although this volume is small, the length of time that the neutrino bombardment has existed will more than compensate — provided, of course, that the trapped argon cannot escape. If we assume that we have recovered (say) 28 cubic metres of rock salt buried for 3 million years (a typical figure), then the argon count should be about 30 million atoms — not a large number, but possible to extract as radioactive gas if the salt is flushed with absolutely pure water. These are rough calculations but they do show that it might be possible to recover trapped argon from the fossilized salt beds that lie beneath the Earth's surface. Note that this system would measure the accumulated argon and give a total integrated neutrino dose, but not a very accurate dose *rate*.

However, if we exploited the oceans we would obtain the opposite effect: we could count the high energy flashes produced as solar neutrinos interacted with the salt solution, and *also* be able to count the gamma flashes from the antineutrinos reacting with the hydrogen protons making up the water content. Both types of event can be distinguished from each other on a pulse-count basis. But we cannot catch the released argon. I would suggest that use of both methods — using both rock salt and ocean — would give the total results.

The 'ocean' neutrino detector would therefore consist of hundreds of scintillator counters and photomultipliers placed in very deep water — 1,500 to 2,000 metres — and connected up to form a three-dimensional grid. If we chose a pilot scheme 10 metres long, 10 metres wide and 10 metres deep, then this would enclose 1,000 cubic metres: if we placed photo counters at 1 metre intervals we would require 1,000 units. The sensitivity of seawater is less than that of carbon tetrachloride, and 1,000 cubic metres (1 million litres) of it is roughly half as sensitive as 450,000 litres of cleaning fluid. But it is cheap, and if the 'pilot' scheme registered (say) five events per year it would tend to confirm the experiments done with cleaning fluid.

It is possible that there are places elsewhere in the Solar System where there are deposits of chloride which may be used as the basis of neutrino detection. For example, Io, the innermost Galilean satellite of Jupiter, appears to have extensive deposits of sodium compounds (believed to be principally salt) and, if this is so, solar neutrino bombardment is producing argon there. The escape velocity of Io is very low, so that it is likely that a stream of argon atoms is escaping from the satellite: these particles might be detectable by sensitive mass spectrometers, and if an orbiting spacecraft is ever sent to Io it should look for traces of this gas.

The conversion of chlorides to argon may also be an explanation for the high argon content in the atmosphere of Venus. This was recorded by the American Pioneer probes and by the Russian Venera 11 spacecraft which landed on the planet in December 1978. The orbit of Venus is 40% closer to the Sun than that of Earth, and as a consequence Venus receives about 2.5 times as many neutrinos. If the atmosphere and rocks contain chlorides, than gaseous argon will be the result of the neutrino bombardment. Being a heavy gas it will be retained by the gravitational field of the planet. Venus may therefore be a 'natural neutrino telescope' or detector awaiting the right keys to unlock the results.

The Ghost of Gravity

Our other future astronomy is the examination of space for signs of gravitational disturbance. By this I do not mean orbital perturbation of the Earth or of the planets due to mutual gravitational fields; I mean the enormous gravitational pulses which ring through space when a giant star collapses.

In what we might narrowly term 'ordinary space' the gravitational force is very weak — about 10^{39} times weaker than electromagnetic forces. We can appreciate this by comparing the gravitational field of the Earth with the electrical fields produced in thunderstorms. The whole mass of Earth is required to raise a field of several billion volts, and the resultant flash of lightning can be destructively forceful. Detecting such powerful electrical fields is easy — listening to the 'atmospherics' or 'crackle' produced by a medium-wave radio receiver on a warm thundery summer day is ample evidence.

Detection of gravitational waves is not so easy.

If mass is required to generate a static gravitational field (as in the case of Earth and the planets) then a *changing* mass will generate a *changing* field. One of the greatest changes we can presently contemplate is the crushing collapse of a star of 15 solar masses in a supernova outburst. The release of energy is considerable. Not only is a *mass* change involved, but also a *diameter* change. The star shrinks from a red supergiant of about 1,000 solar diameters (according to latest estimates) down to 15km or possibly less. This represents a ratio of about 10^8. If the mass had remained the same, the gravitational field would have increased by 10^{16}, but the mass goes down by a factor of 10 in the explosion and collapse, and this modifies the final answer to 10^{15}. Therefore, during its collapse, a giant star produces a gravitational field change equal to about 1 million billion times the strength of the gravitational field of our Sun! It is the equivalent of the field of a small galaxy. This is for a Type II supernova; a Type I is described on page 200.

The pulse, once produced, will travel throughout the breadth of our Galaxy and possibly be detected as a 'ring' which may last a few minutes. The initial collapse will take about a second, but the star core will oscillate violently with a frequency of a few kilohertz (i.e., a few thousand times per second). In air this would be a musical note similar to a piano string or a bell chime which sounds and gradually dies away.

It was these oscillations that Joseph Weber of the Princeton Institute of Advanced Study set out to detect. He used the principle of resonance in constructing the detecting apparatus. If we place two pianos in close proximity and strike a key (say middle C) fairly hard with the pedal depressed, then (provided its pedal has also been depressed and it is accurately tuned) the second piano will emit middle C also, even though the keys have not been touched. The string of the second piano is vibrating in sympathy, or resonating, with that of the first.

The experiment being carried out by Weber involved 'tuning' a pair of aluminium cylinders such that they resonated or 'sang' when struck by a gravitational wave. Weber chose aluminium because it is cheap and plentiful (compared with other metals) and is nonmagnetic. The cylinders were very carefully suspended from wires which in turn were mounted from a shock-isolated frame and the complete assembly housed in a vacuum; these extreme precautions were taken to isolate causes of cylinder movement due to sources other than gravitational waves. The movements of people, vehicles and even trees in the wind had to be accounted for. Earthquake vibrations were monitored by means of a very sensitive seismograph. Other unwanted signals were filtered out prior to recording.

The movements of the cylinder deemed appropriate to gravitational waves were picked up by a row of quartz piezoelectric crystals strapped round the waist of the cylinder. These converted the cylinder vibrations into minute electric currents which were then amplified, filtered and recorded, together with a timing code.

Two installations have been used by Weber, one at the Argonne National Laboratory in Chicago and the other at the University of Maryland College park some 1,000km away. Each cylinder installation had its own recording facilities, but both shared a common timing system which enabled a computer to look for simultaneous events (to be more precise, they would be separated by 30 microseconds, the time it takes to travel 1,000km at the speed of light). This correlation was necessary because, in spite of the elaborate precautions, a considerable amount of residual vibration (traffic, Earth movements, etc.) broke through. Correlation cancelled this and left events which looked remarkably like the elusive gravitational waves.

Unfortunately, the events appeared to be too frequent for some critics. The pulses are in bursts and come from different parts of the Galaxy, with a preference for emanating from the direction, apparently, of the *centre* of the Galaxy. The average rate of occurrence is one per day, and this raises problems if based on our ideas of the frequency with which supernova outbursts occur and are actually observed in various galaxies (including our own). Originally supernovae were thought to occur once every 300 years, this was reduced to every 150 years and now they are thought to occur as frequently as every 40 years (on average) — but see page 136. But if supernova outbursts occurred at the rate of one per day then the Galaxy — thought to contain about 180 billion stars — would be devoid of all stars within about 500 million years (the Galaxy is thought to be 15 *billion* years old).

This excess of gravitational-wave 'events' has caused some critics to be very sceptical of the evidence, even to the extent of suggesting that Weber was not treating his results in an unbiased manner. Weber's reply is that it is the computer that is picking out the correlated events and results are therefore free from bias or prejudice.

Matters are not improved when other experimenters, who claim to have more sensitive apparatus, do not appear to achieve the same results. Indeed, they claim they have failed to record *any* pulses that might be due to gravitational waves, and at present matters seem to have reached an impasse.

There are two ways around the problem. One is to duplicate Weber's apparatus *exactly* and use his filters, detectors, cylinder suspension and computer correlation programme. Not only will an independent opinion be obtained, but completely independent results. The second step in this first approach would be to use Weber's timing signals as a time base for the recording. In this way the combined results of two groups in different parts of the world could be assessed by an unbiased party.

The second approach is to consider *other* processes which might produce gravitational waves besides the outburst of a supernova. I would suggest an alternative process which does occur far more frequently than supernova outbursts. I refer to the so called 'X-ray bursters' or 'rapid fire X-ray sources' which were discovered in 1974 with the launching of Ariel 5. These powerful outbursts are now referred to as 'X-ray novae', and involve the mechanism of mass transfer between binary sources.

If we consider that there are pairs of stars where the mass transfer is large (say 0.1 solar mass) and sudden (a few seconds), then there is no reason why gravitational waves cannot be produced if the companion star is of the right proportions — either a massive neutron star or a black hole. If the other star is a large blue supergiant (30 to 50 solar masses) then a 0.1 solar mass transfer may take place several times before the blue giant is exhausted.

We can further calculate how far the material would have to fall towards its partner by using the analogy of the collapse of a star from a red supergiant to a neutron star. In the case of an X-ray 'transient recurrent supernova' the mass transferred is, as we have mentioned, estimated as 0.1 solar mass. Therefore, to produce the same gravitational energy as the collapse of a red giant of solar mass, this small mass has to move through a distance of $\sqrt{10}$ or about 3.2 times the radius of such a star. This corresponds to a separation of about 12.5 AU.

In order that mass transfer may take place the companion must come within the Roche Limit of the blue supergiant main star. This demands that the star be about 5 AU in radius — about the right size for a blue supergiant.

It is, therefore, possible that Weber's gravitational waves are being caused not by supernovae at all but by transfer of sizeable masses from a blue supergiant to a black hole or dense neutron star. The procedure may be repeated several times (maybe 200 or more) before the exhaustion of supergiant material, or the actual explosion of the supergiant as a true supernova. This, I think, offers a way out of the Weber dilemma, and could be verified by careful counts of X-ray novae for signs of an X-ray supernova.

Detectors in Space

In an earlier section of this chapter we mentioned the problems of counting solar antineutrinos — none are produced by the solar fusion reaction; and an Earth-bound antineutrino telescope has to contend with the terrestrial background antineutrino 'noise'.

As a star moves towards supernova, and just prior to the explosion, it produces antineutrinos in quantity. At a final peak (when oxygen is being converted to silicon), just before iron generation and collapse, the neutrino-antineutrino 'brightness' of the doomed star is about 6×10^{11} that of the Sun. Therefore it may be possible to use *antineutrino* detectors in space where the terrestrial background would be very low and the solar background negligible. The terrestrial limit sets a detector sensitivity limit; it appears that a supernova at 1.5 kiloparsecs would be detectable by a satellite in geosynchronous orbit. A satellite in higher orbit would give greater sensitivity, and an antineutrino counter on the Moon *might* do better still, although we would need to know the actual lunar background count.

S. Sofia and W. M. Sparkes of Goddard Space Flight Centre in Maryland, and A.S. Endal of Louisiana University, propose just such an experiment in geosynchronous orbit (*Nature* for 8 February 1979). They do point out, however, that 'the technical difficulties involved in designing a suitably sensitive detector [for a spacecraft] should not be underestimated'.

Detection of these pre-outburst increases would be very useful as the *first* major increase (according to Sofia, Sparkes, and Endal) of antineutrinos would occur 4 months before the collapse, a second peak 1 month, a third approximately 14 days, a fourth 10 days, and a fifth 4 days prior. Then the final collapse would produce a particle flood which would go 'off the clock' at over 10^{12} (possibly approaching 10^{16}) times the solar neutrino output. The detectors suggested would therefore easily 'see' a supernova event *on the other side of the Galaxy* and also 'see' the pre-event sequence and its singular and unique fingerprint.

Solid-state crystal detectors heavily shielded in lead would be required — hence the 'technical difficulties'. Such detectors could be of caesium iodide or similar halogen salt.

The problems of putting Weber's cylinders into space are not so severe, and it is possible that this may be done in the course of the next ten years. Placed in a geosynchronous orbit, two or more cylinders would give clear hints as to the direction and magnitude of the gravitational wavefronts. The problems of earthquakes would be eliminated, but there would be a need for compact, solid construction of the spacecraft. In particular, the craft would have to be at a uniform temperature. 'Creaks and groans' due to structural distortion with temperature fluctuations would be intolerable and show up as 'noise' on the recorded and processed data. A total integrated system for antineutrino *and* gravitational detection would provide a very comprehensive and complementary system of supernova detection.

It would truly mark a major advance in the new astronomies.

Opening Frontiers — The New Cosmologies

Cosmology is a great big gamble — winner takes all — on a series of ideas and theories; you back a theory as you might back a horse, except there are no rewards for a 'place'. And unlike the case of a horse race, we did not see the beginning and we will never see the end. Our view is that of a bystander who is allowed one peep at the race after it has started and who must decide and judge the winner on the strength of that single scanning glance.

In the true manner of horse racing, the established favourites of known form and pedigree are there, and at least one has a good chance of winning it. In this race the favourites were originally the 'Big Bang' and 'Steady State' theories, and as the prospects changed so did the odds. Also in the true manner of horse racing, there are outsiders and 'unknowns', horses who are not at all well known and whose chances are consequently rated low. In this class we have the 'Oscillating Universe', and the new challenger 'Heat Death'. This last has a good pedigree and could easily become the winner of the Cosmological Derby.

I will attempt to explain the newcomer; the well established 'Big Bang', 'Oscillating Universe' and 'Steady State' theories I will not deal with in detail as they have been adequately described over the last twenty years. It is sufficient to say that the Big Bang has continued to be a viable theory over this period and is probably still the favourite.

The Unstable Eternity of Heat Death

This theory was recently advocated in *Nature* for 30 November 1978 by John Barrow of Oxford and Frank Tipler of Berkeley, California. They both assume that the Universe was born in something like the Big Bang — but then they become concerned about the way it will die.

And here we encounter our friend entropy once again. We discussed the concept earlier (page 183), and showed its relation with the surface area of a black-hole event horizon. Here we consider entropy as a state of *diminishing change* — if the entropy rises to its absolute maximum, the Universe is in a state of unchanging death.

This was first outlined by von Helmholtz in 1854 and amplified by Eddington in 1931. The latter visualized the Universe growing larger and thinner, the matter slowly changing into radiation, which moved to longer and longer wavelengths. Eddington foresaw the Universe as doubling its radius every 1,500 million years and expanding in this way forever. In terms of temperature this means that the background drops to 2.1 K in 1,500 million years, 1.0 K in 3,000 million, 0.7 K in 4,500 million, 0.5 K in 6,000 million, 0.35 K in 7,500 million, and so on. This illustrates the slow decrease in energy levels as time passes.

Barrow and Tipler have modified and extended this idea by including the modern concepts of quantum theory and black holes. They consider that all matter will drift to the black-hole state. Stars of small mass above Chandrasekhar's Limit will be the first to attain this condition, but larger masses will eventually follow, and they quote a mass of 10^{15} solar masses (approximately 5,000 galaxies) as taking about 10^{19} years (10 million million

216

million) to do so. Material may be radiated from these holes but it will gather together to form stars which will eventually in turn become black holes; that is, there is a grand progression through many cycles of infall and evaporation. This movement is cyclic, but progresses inexorably toward a state of supermassive black holes, debris of 'dead' matter (black dwarfs, dead planets), elementary particles (cool quarks) and low-temperature background radiation. Applying quantum mechanics to this 'mess of potage' Barrow and Tipler find that the black holes (even though massive) will eventually evaporate and leave *nothing* but elementary particles and radiation.

Here a further novelty is introduced, and this involves the entropy of the gravitational fields which will permeate the Universe. Although they will be strengthened and weakened cyclically as the material is processed and reprocessed, the entropy of these fields will also increase overall. In other words, the spacetime fabric of Einstein and Minkowski is breaking up and tearing apart. When this happens there is no question of the elementary particles reforming as new matter, for the spacetime fabric on which the new matter would be woven has disintegrated — and, long before that, it will have become very 'patchy'. This means that a finite, ever-expanding Universe as visualized by Eddington is impossible; for such a Universe — even if only of pure low-temperature radiation — will exist only for a finite physical time.

Tipler and Barrow finally point out that, as a rough approximation, only a finite number of changes will occur in the future: the Heat Death puts an end to change! There will be no matter, no radiation, no background 'temperature', no gravitational spacetime fields — nothing.

Strange and eerie though this far future may seem, it is perfectly permissible in quantum theory. In Chapter 11 we briefly considered Pauli's concept of a sea of virtual particles as applied to black holes. In the general case he visualized a vast 'sea' of particles and antiparticles which, when taken together, formed space. What Barrow and Tipler have done is to point out that the Heat Death represents the ultimate stillness where every particle is nulled or cancelled by an antiparticle, and where there is no movement of particles or between particles. The particle sea is utterly at rest; the time by which any movement could be judged is also absent.

A future so drastically complete is semimystical, and begs the question: 'How were the waters of Pauli's sea of particles disturbed in the first place, such that the Big Bang occurred, the Universe was born, and ultimately we came into being?'

It is a question we cannot presently answer — and I suspect that we may *never* be able to answer it.

Index

Numbers in italic type indicate illustrations

219

220

222